MW00814511

146
Structure and Bonding

Series Editor: D. M. P. Mingos

Editorial Board:
F. A. Armstrong · P. Day · X. Duan · L. H. Gade
K. R. Poeppelmeier · G. Parkin · J.-P. Sauvage ·
M. Takano

For further volumes:
http://www.springer.com/series/430

Structure and Bonding

Series Editor: D. M. P. Mingos

Recently Published and Forthcoming Volumes

Electron Density and Chemical Bonding I
Volume Editor: Dietmar Stalke
Vol. 146, 2012

Structure-Property Relationships in Non-Linear Optical Crystals II
Volume Editors: Xin-Tao Wu, Ling Chen
Vol. 145, 2012

Structure-Property Relationships in Non-Linear Optical Crystals I
Volume Editors: Xin-Tao Wu, Ling Chen
Vol. 144, 2012

Molecular Electronic Structures of Transition Metal Complexes II
Volume Editors: D. M. P. Mingos, Peter Day, Jens Peder Dahl
Vol. 143, 2012

Molecular Electronic Structures of Transition Metal Complexes I
Volume Editors: D. M. P. Mingos, Peter Day, Jens Peder Dahl
Vol. 142, 2012

Fuel Cells and Hydrogen Storage
Volume Editors: Andrew Bocarsly, D. M. P. Mingos
Vol. 141, 2011

Zintl Ions
Principles and Recent Developments
Volume Editor: Thomas F. Fässler
Vol. 140, 2011

Zintl Phases
Principles and Recent Developments
Volume Editor: Thomas F. Fässler
Vol. 139, 2011

Inorganic 3D Structures
Volume Editor: Angel Vegas
Vol. 138, 2011

Molecular Catalysis of Rare-Earth Elements
Volume Editor: Peter W. Roesky
Vol. 137, 2010

Metal-Metal Bonding
Volume Editor: Gerard Parkin
Vol. 136, 2010

Functional Phthalocyanine Molecular Materials
Volume Editor: Jianzhuang Jiang
Vol. 135, 2010

Data Mining in Crystallography
Volume Editors: Hofmann, D. W. M., Kuleshova, L. N.
Vol. 134, 2010

Controlled Assembly and Modification of Inorganic Systems
Volume Editor: Wu, X.- T.
Vol. 133, 2009

Molecular Networks
Volume Editor: Hosseini, M. W.
Vol. 132, 2009

Molecular Thermodynamics of Complex Systems
Volume Editors: Lu, X., Hu, Y.
Vol. 131, 2009

Contemporary Metal Boron Chemistry I
Volume Editors: Marder, T. B., Lin, Z.
Vol. 130, 2008

Recognition of Anions
Volume Editor: Vilar, R.
Vol. 129, 2008

Liquid Crystalline Functional Assemblies and Their Supramolecular Structures
Volume Editor: Kato, T.
Vol. 128, 2008

Organometallic and Coordination Chemistry of the Actinides
Volume Editor: Albrecht-Schmitt, T. E.
Vol. 127, 2008

Electron Density and Chemical Bonding I

Experimental Charge Density Studies

Volume Editor: Dietmar Stalke

With contributions by

L.J. Farrugia · U. Flierler · C. Hauf · V. Herz · B.B. Iversen · D. Leusser · P. Macchi · A. Østergaard Madsen · J. Overgaard · W. Scherer · D. Stalke

Editor
Dietmar Stalke
Universität Göttingen
Institut für Anorganische Chemie
Tammannstraße 4
Göttingen
Germany

ISSN 0081-5993 ISSN 1616-8550 (electronic)
ISBN 978-3-642-30801-7 ISBN 978-3-642-30802-4 (eBook)
DOI 10.1007/978-3-642-30802-4
Springer Heidelberg New York Dordrecht London

Library of Congress Control Number: 2012939135

© Springer-Verlag Berlin Heidelberg 2012
This work is subject to copyright. All rights are reserved by the Publisher, whether the whole or part of the material is concerned, specifically the rights of translation, reprinting, reuse of illustrations, recitation, broadcasting, reproduction on microfilms or in any other physical way, and transmission or information storage and retrieval, electronic adaptation, computer software, or by similar or dissimilar methodology now known or hereafter developed. Exempted from this legal reservation are brief excerpts in connection with reviews or scholarly analysis or material supplied specifically for the purpose of being entered and executed on a computer system, for exclusive use by the purchaser of the work. Duplication of this publication or parts thereof is permitted only under the provisions of the Copyright Law of the Publisher's location, in its current version, and permission for use must always be obtained from Springer. Permissions for use may be obtained through RightsLink at the Copyright Clearance Center. Violations are liable to prosecution under the respective Copyright Law.
The use of general descriptive names, registered names, trademarks, service marks, etc. in this publication does not imply, even in the absence of a specific statement, that such names are exempt from the relevant protective laws and regulations and therefore free for general use.
While the advice and information in this book are believed to be true and accurate at the date of publication, neither the authors nor the editors nor the publisher can accept any legal responsibility for any errors or omissions that may be made. The publisher makes no warranty, express or implied, with respect to the material contained herein.

Printed on acid-free paper

Springer is part of Springer Science+Business Media (www.springer.com)

Series Editor

Prof. D. Michael P. Mingos

Inorganic Chemistry Laboratory
Oxford University
South Parks Road
Oxford OX1 3QR, UK
michael.mingos@st-edmund-hall.oxford.ac.uk

Volume Editor

Dietmar Stalke
Universität Göttingen
Institut für Anorganische Chemie
Tammannstraße 4
Göttingen
Germany

Editorial Board

Prof. Fraser Andrew Armstrong

Department of Chemistry
Oxford University
Oxford OX1 3QR
UK

Prof. Peter Day

Director and Fullerian Professor
of Chemistry
The Royal Institution of Great Britain
21 Albermarle Street
London W1X 4BS, UK
pday@ri.ac.uk

Prof. Xue Duan

Director
State Key Laboratory
of Chemical Resource Engineering
Beijing University of Chemical Technology
15 Bei San Huan Dong Lu
Beijing 100029, P.R. China
duanx@mail.buct.edu.cn

Prof. Lutz H. Gade

Anorganisch-Chemisches Institut
Universität Heidelberg
Im Neuenheimer Feld 270
69120 Heidelberg, Germany
lutz.gade@uni-hd.de

Prof. Dr. Kenneth R. Poeppelmeier

Department of Chemistry
Northwestern University
2145 Sheridan Road
Evanston, IL 60208-3133
USA
krp@northwestern.edu

Prof. Gerard Parkin

Department of Chemistry (Box 3115)
Columbia University
3000 Broadway
New York, New York 10027, USA
parkin@columbia.edu

Prof. Jean-Pierre Sauvage

Faculté de Chimie
Laboratoires de Chimie
Organo-Minérale
Université Louis Pasteur
4, rue Blaise Pascal
67070 Strasbourg Cedex, France
sauvage@chimie.u-strasbg.fr

Prof. Mikio Takano

Institute for Integrated Cell-Material
Sciences (iCeMS)
Kyoto University
Yoshida Ushinomiya-cho
Sakyo-ku
Kyoto 606-8501
Japan

Structure and Bonding
Also Available Electronically

Structure and Bonding is included in Springer's eBook package *Chemistry and Materials Science*. If a library does not opt for the whole package the book series may be bought on a subscription basis. Also, all back volumes are available electronically.

For all customers who have a standing order to the print version of *Structure and Bonding*, we offer the electronic version via SpringerLink free of charge.

If you do not have access, you can still view the table of contents of each volume and the abstract of each article by going to the SpringerLink homepage, clicking on "Chemistry and Materials Science," under Subject Collection, then "Book Series," under Content Type and finally by selecting *Structure and Bonding*.

You will find information about the

- Editorial Board
- Aims and Scope
- Instructions for Authors
- Sample Contribution

at springer.com using the search function by typing in *Structure and Bonding*.

Color figures are published in full color in the electronic version on SpringerLink.

Aims and Scope

The series *Structure and Bonding* publishes critical reviews on topics of research concerned with chemical structure and bonding. The scope of the series spans the entire Periodic Table and addresses structure and bonding issues associated with all of the elements. It also focuses attention on new and developing areas of modern structural and theoretical chemistry such as nanostructures, molecular electronics, designed molecular solids, surfaces, metal clusters and supramolecular structures. Physical and spectroscopic techniques used to determine, examine and model structures fall within the purview of *Structure and Bonding* to the extent that the focus is on the scientific results obtained and not on specialist information concerning the techniques themselves. Issues associated with the development of bonding models and generalizations that illuminate the reactivity pathways and rates of chemical processes are also relevant.

The individual volumes in the series are thematic. The goal of each volume is to give the reader, whether at a university or in industry, a comprehensive overview of an area where new insights are emerging that are of interest to a larger scientific audience. Thus each review within the volume critically surveys one aspect of that topic and places it within the context of the volume as a whole. The most significant developments of the last 5 to 10 years should be presented using selected examples to illustrate the principles discussed. A description of the physical basis of the experimental techniques that have been used to provide the primary data may also be appropriate, if it has not been covered in detail elsewhere. The coverage need not be exhaustive in data, but should rather be conceptual, concentrating on the new principles being developed that will allow the reader, who is not a specialist in the area covered, to understand the data presented. Discussion of possible future research directions in the area is welcomed.

Review articles for the individual volumes are invited by the volume editors.

In references *Structure and Bonding* is abbreviated *Struct Bond* and is cited as a journal.

Impact Factor in 2010: 4.659; Section "Chemistry, Inorganic & Nuclear": Rank 4 of 43; Section "Chemistry, Physical": Rank 25 of 127

Preface

Ever since Wilhelm Conrad Röntgen discovered X-rays in Würzburg and von Laue, Ewald, Friedrich, Knipping, and Bragg developed analytical methods to interpret X-ray diffraction patterns a hundred years ago in Göttingen, experimental single crystal structure analysis from diffraction data has become the most powerful analytical tool used to elucidate unequivocally the three dimensional structures of solid matter. Since that time the technique has contributed to many Nobel prizes in chemistry, physics and life sciences. From a knowledge of the connectivities at the atomic level and the arrangement in the solid phase many properties, both at the molecular and macroscopic scale, can be deduced. Currently X-ray sources in research laboratories and large facilities are getting more and more powerful, detectors more and more sensitive and crystallographic programs more and more sophisticated, so that the method continues to flourish and has been successfully applied to problems of increasing complexity. However, the most basic concept, the chemical bond, is still vigorously discussed and at times debated since its essential features were introduced by Lewis and put on a quantum mechanical basis by Pauling and Mulliken. Still there is room for interpretation, because single crystal structural analyses based on the independent atom model only provides the positions of the centroids of the atoms and the distances between the atoms. In the electron density maps there are no lines or dashes defining or even indicating the chemical bond and the nature of the bonding remains a matter of interpretation based on a bonding model. Hence the anecdote that *a bond is where the chemist draws the line* remains valid to a certain extent. Most of our understanding of the chemical bond is still deduced from the distances and angles, which are determined as a result of the crystallographic analysis, and represents a conclusion from a logical causal chain. For example, chemical intuition suggests that a short bond is a strong bond and possibly it has multiple bond character making it difficult to cleave, but nevertheless remaining reactive towards polar reagents. Hence we employ knowledge compiled from databases and statements of masterminds to indirectly deduce the nature of the bond. However, increasingly it has been recognised that there are exceptions to these simple generalisations, e.g. increasing 'evidence' that short bonds are not necessarily strong or multiple bonds and multiple bonds may at times be longer than single bonds. Furthermore, there may be features of bonds which we cannot explain using current concepts.

This explains the increasing interest in looking at the bond directly and not by means of deductions and statistics. The Hansen and Coppens partitioning of the electron density in the aspherical atom model paved the way to describe the interference of X-rays with electrons from high resolution diffraction data more adequately. This density may then be analysed by Bader's Quantum Theory of Atoms in Molecules, directly providing more resilient and physically meaningful descriptions of chemical bonding than inferences from bond distances and angles.

The following collection of articles in Volumes 146 and 147 of the Structure and Bonding Series provides a state-of-the-art overview of the capacity of the method written by leaders in the field, which we hope will motivate more scientists to take advantage of the general approach.

Göttingen
April 2012

Dietmar Stalke

Contents

Struct Bond (2012) 146: 1–20
DOI: 10.1007/430_2012_80
© Springer-Verlag Berlin Heidelberg 2012
Published online: 17 April 2012

More than Just Distances from Electron Density Studies

U. Flierler and D. Stalke

Abstract This chapter provides a short synopsis of the basics of electron density (ED) investigations. The two predominant approaches for modeling electron density distributions (EDDs) are explained and compared. Various possible interpretations of the resulting EDD are presented and their potential translations into chemical concepts are depicted.

Keywords Charge density determination · Independent atom model · Multipole model · Quantum theory of atoms in molecules · Structure determination

Contents

U. Flierler and D. Stalke (✉)
Georg-August-Universität Göttingen, Institut für Anorganische Chemie, Tammannstraße 4, 37077 Göttingen, Germany
e-mail: dstalke@chemie.uni-goettingen.de

1 Introduction

Synthetic chemists – no matter in which field of chemistry they are working – are always searching for new, improved compounds. The design of tailor-made reagents, precisely adapted to specific applications, is one of the fundamental objectives in chemistry. While the exact synthetic target might differ, the overarching issue is the same: selective replacement of individual atoms or residues within a molecule to fine-tune its properties and by these means its reactivity or functionality. However, it is still not always predictable which alterations cause the desired effects. The design of new target molecules is often based on a long process of trial and error. Therefore, a proper understanding of the relation between structure and chemical or physical properties of a compound is vital for the improvement of this development process. Electron density (ED) investigations provide an ideal tool for understanding these interactions [1–9]. Already in 1932, Pauling [10] proposed an interrelation between the structure of a compound and its properties:

> The properties of a compound depend on two main factors, the nature of the bonds between the atoms, and the nature of the atomic arrangement. [...] The satisfactory description of the atomic arrangement in a crystal or molecule necessitates the complete determination of the position of the atoms relative to one another.

The claimed determination of the nuclear positions was already feasible at that time via X-ray structure determination [11]. The measurement of the intensity and the spatial distribution of the diffracted X-rays yields – via integration – the structure factors F_{obs}. Theoretically, passing through a Fourier transformation, the ED $\rho(\mathbf{r})$ can be calculated from these structure factors [12]. However, this direct evaluation causes some problems due to experimental shortcomings:

- The observed structure factors are affected by experimental errors. Even if these errors are corrected by different mathematical routines, they are nevertheless inherent in the data.
- As only a finite number of reflections can be collected, Fourier truncation errors occur.
- The phase information is lost during the measurement. This is due to the fact that the measured intensities are proportional to the *squared* structure factors.

Therefore, the ED in the crystal can actually not directly be calculated from the observed structure factors and thus has to be modeled. In practice, the Fourier transformation can be avoided, because quantum mechanics facilitates the construction of a mathematical model of the ED in a crystal. If the arrangement of the atomic nuclei in the crystal lattice is known, the structure factors can be calculated from a parameterized model. These calculated structure factors F_{calc} are then compared to the observed structure factors F_{obs}. By optimizing the parameters of the calculated model, F_{calc} is adjusted to F_{obs} in a least-squares refinement. This way, the above-mentioned practical problems can be solved. Experimental errors are typically minimized by the least-square refinement. The models do not suffer from truncation errors and the phase is reconstructed from the model itself. In addition, the experimental density is inherently dynamic but some methods (especially the least-squares refinement of atom expanded density) allow to extract a static ED. This is important for comparative purposes. Thus, from the diffraction experiment, the electron density distribution (EDD) in the unit cell of a single crystal is available.

2 Model Generation

2.1 The Independent Atom Model

The independent atom model (IAM) is one approach to derive the calculated structure factors F_{calc}. This spherical atom approximation was and is used for almost all X-ray standard structure determinations. Within this approach, the density is described as a superposition of spherical atomic densities that – via a Fourier transformation – gives the spherical atomic scattering factors [13].

From a chemical understanding, the density is accumulated around the nuclei of the atoms. Therefore, the maxima of the EDD in the unit cell can be interpreted to be the nuclear positions of the molecule. The assumption of a spherical distribution of the density around these nuclei is adequate to obtain a reasonable model of the structure, i.e., the connectivities between the atoms. However, bonding effects and lone pairs are not considered (cf. Fig. 1).

There are up to nine parameters per atom that can be altered to adjust F_{calc} to F_{obs}: three positional parameters (x, y, and z) and six anisotropic displacement parameters (U_{ij}) for non-hydrogen atoms or one isotropic displacement parameter (U_{iso}) usually used only for hydrogen atoms, respectively. These parameters describe the assumed spherical atoms.

The distances between these spherical atoms – the bond lengths – and the bond angles are utilized to derive information on the type of interactions between them. However, the conclusions drawn can also be misleading, suggesting a bonding that does not hold on a closer inspection. For example, the suggestive correlation that a short bond implies a strong bond was falsified already 10 years ago [14]. Additionally, the recently synthesized and derivatized E–E multiple bonds (E = heavier Group 14 elements) clearly put the established bond length/reactivity correlation into a new perspective [15, 16].

Already Pauling [10] warned, that

> ... There is, of course, a close relation between atomic arrangement and bond type. But [...] it is by no means always possible to deduce the bond type from a knowledge of the atomic arrangement.

However, the theoretical background for a direct examination of the bonding was known as early as X-ray diffraction itself. As X-rays are scattered much stronger by the *electrons* than by the *nuclei* of an atom, the result obtained from a diffraction experiment depends basically on the EDD in a molecule. Especially for light atoms, Debye [17] proposed already in 1915, that “it should be possible to

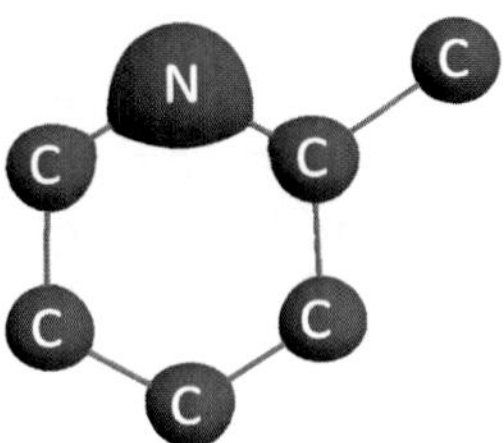

Fig. 1 ED $\rho(\mathbf{r})$ of a coordinated picoline-ring modeled with the IAM approach (taken from [8])

determine the arrangement of the electrons in the atoms." Nevertheless, the information contained was experimentally not accessible for almost half a century. This can be imputed, on one hand, to the inaccuracy of the data and, on the other hand, to the great success of the spherical IAM. As the EDD of an atom is distorted because of its interactions with other atoms, information on these interactions can only be obtained, if the aspherical density is described.

2.2 *The Multipole Model*

In the 1960s, Dawson [18], Stewart [19–22], and Hirshfeld [23, 24] started discussing the application of aspherical atoms to be able to describe the bonding contributions, and the atom-centered finite multipole expansion (multipole model, MM) by Hansen and Coppens [25] finally leveraged the aspherical description of the density. Within this approach, the atomic density $\rho(\mathbf{r})$ is divided into three components (1): the spherical core density $\rho_c(\mathbf{r})$, the spherical valence density $\rho_v(\kappa\,\mathbf{r})$, and the aspherical valence density $\rho_d(\kappa'\mathbf{r})$.

$$\rho_{\text{atom}}(\mathbf{r}) = P_c\rho_c(\mathbf{r}) + P_v\kappa^3\rho_v(\kappa\mathbf{r}) + \rho_d(\kappa'\mathbf{r})$$

with

$$\rho_d(\kappa'\mathbf{r}) = \sum_{l=0}^{l\max} \kappa'^3 R_l(\kappa'\mathbf{r}) \sum_{m=0}^{l} P_{lm\pm} d_{lm\pm}(\theta, \Phi) \tag{1}$$

κ and κ' represent radial scaling parameters that allow for an expansion or a contraction of the spherical valence density and the aspherical valence density, respectively. The core and the spherical valence densities, $\rho_c(\mathbf{r})$ and $\rho_v(\kappa\,\mathbf{r})$, are calculated from Hartree–Fock (HF) [26] or relativistic HF [27–29] atomic wave functions. Their values are tabulated in the literature [30]. The radial functions of the aspherical deformations density, $\rho_d(\kappa'\mathbf{r})$, are described by spherical harmonics, the multipoles, which provide a much more flexible model for the description of the measured density.

Within the multipole model, this is achieved by the charge density parameters, as P_v, κ and κ' which are, in addition to the nine conventional parameters described above, optimized in a least-squares refinement based on the measured structure factors. This results in up to 36 parameters for each non-hydrogen atom.

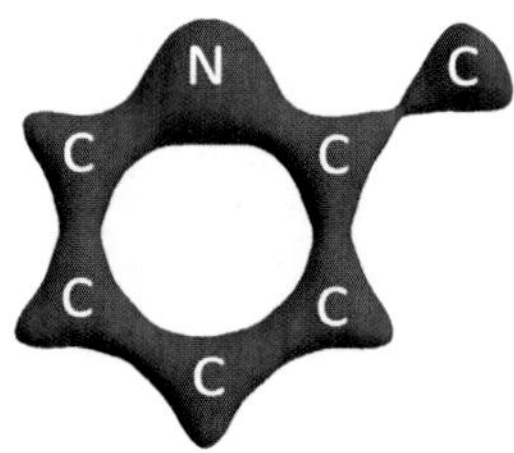

Fig. 2 ED $\rho(\mathbf{r})$ of a coordinated picoline-ring modeled with MM approach (taken from [8])

Thus, the multipolar refinement offers a convincing advantage over the independent atom refinement: an accurate model that is suitable to describe not only the spherical atom density but also the aspherical density contributions (cf. Fig. 2). Since these aspherical density contributions stem mainly from bonding effects on the atoms, the modeled density will contain information about the interactions between the atoms.

2.3 *Reliability of the Model*

R-Values: During the least-squares refinement, the difference between the observed and the calculated structure factors is minimized. How well this fit matches is expressed by the R-values $R1$ (2) and $wR2$ (3). If the model is refined against F^2, the $wR2$-value is more significant, and the closer the R-values are to zero, the better the match. w_H denotes a weighting scheme.

$$R1 = \frac{\sum_{H} ||F_{\text{obs}}| - |F_{\text{calc}}||}{\sum_{H} |F_{\text{obs}}|} \tag{2}$$

$$wR2 = \sqrt{\frac{\sum_{H} w_H \left(|F_{\text{obs}}|^2 - |F_{\text{calc}}|^2\right)^2}{\sum_{H} w_H |F_{\text{obs}}|^4}} \tag{3}$$

Residual Density: The residual density is another indicator for the quality of the refinement. It represents the difference between the modeled and the observed ED. The smaller the difference is, the better the model describes the molecular density. The most important test to judge on the success of the model and the quality of the fit is to evaluate the residual ED through a Fourier summation ($F_{\text{obs}} - F_{\text{calc}}$). This provides a direct-space representation of the extent to which the model accounts for the observations.

Strongly featured residual density maps are inevitable in the IAM refinement because it models the atoms spherically and would not account for the ED in bonding regions. Therefore, especially the bonding regions show strong positive residual density peaks (Fig. 3b). During the multipole refinement, these residuals

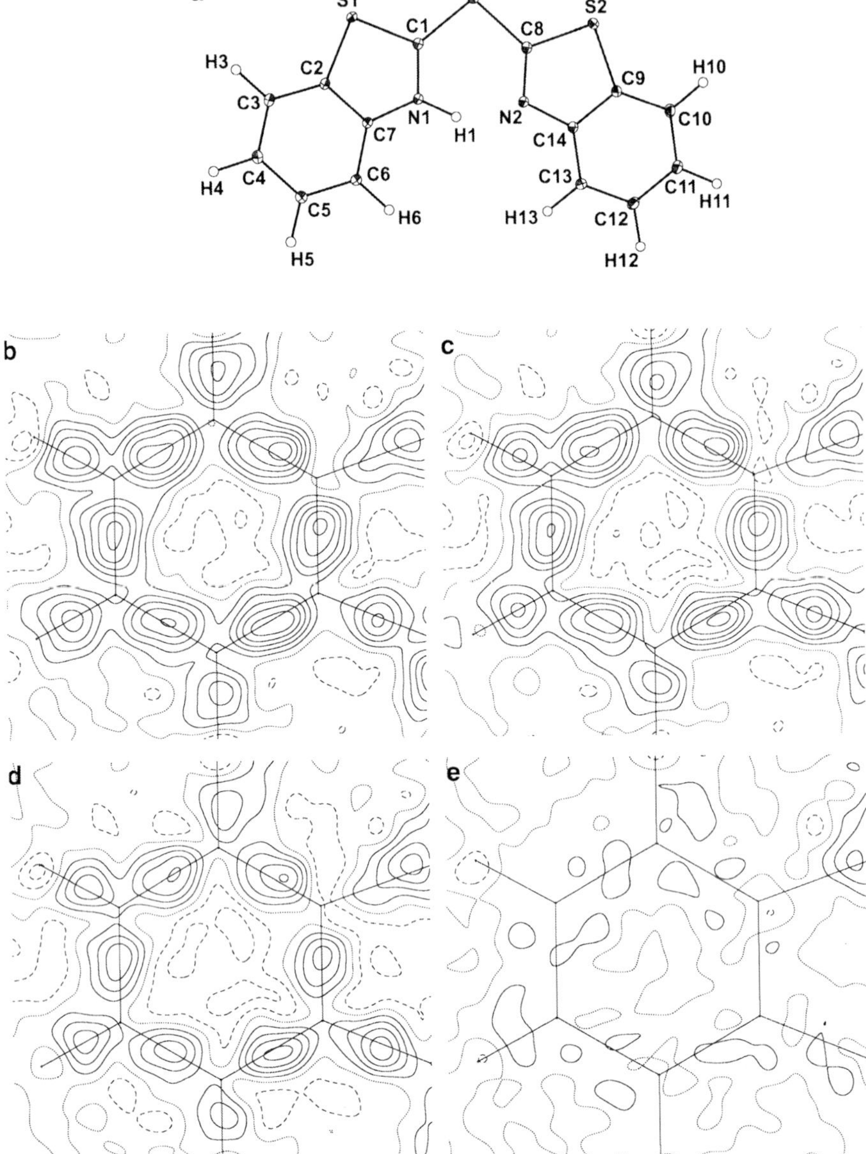

Fig. 3 Residual density maps in the C_6 perimeter plane of $HP(bth)_2$ (**a**) in dependence of the maximum order of the refined multipoles for the ring carbon atoms shown above; (**b**) monopoles, (**c**) dipoles, (**d**) quadrupoles, (**e**) octapoles, positive values of the residual density *solid lines*, negative values *dashed*, zero contour *dotted*, stepsize 0.1 e/Å^3

are accounted for and the resulting residual density map is virtually flat and featureless (Fig. 3e). Figure 3 shows the dependence of the fitted residual density from the maximum order of expansion of refined multipoles. In the case of the sp^2-hybridized carbon atoms of the benzothiazolyl ring, the quadrupoles and first of all the octapole functions are describing the bonding densities in the C_6 perimeter. If any residual density is left, it should be randomly distributed. It can then be assigned to experimental noise. Errors in the model can be concluded from residual density analysis (RDA) in proximity to the core, while residual densities in larger distances to atom positions are indicative of bad data quality.

A flat and featureless residual map is a necessary condition for the adequacy of a model, but it is far from being a sufficient one to judge its physical significance. A quantitative analysis of the residual density distribution is available via the RDA [31].

Goodness of Fit: Next to the weighted R-values, the Goodness of Fit (S, GoF, 4) is another important quality criterion. It shows the relation between the deviation of F_{calc} from F_{obs}, and is thus a measure of the over-determination of refined parameters.

$$S = \sqrt{\frac{\sum\left(w_H\left(F_{\text{obs}}^2 - F_{\text{calc}}^2\right)^2\right)}{(n-p)}} \tag{4}$$

Here, n is the number of reflections, w_H is the weighting scheme applied, and p is the number of parameters.

For a correct structure refinement with an adapted weighting scheme, S should be one. Higher values for the GoF are due to a systematic underestimation of the uncertainties of the reflections at higher Bragg angles. Therefore, the quality of the model seems to be worse than expected from the quality of the data. In addition, the GoF for a multipole refinement is usually higher than one due to the weighting scheme applied. This is owed to the dependence of S from the number of the observations and variables. For high resolution data, this ratio is unfavorable with respect to the calculated S. An adjustment of the weights reduces S but flattens the overall residual densities especially in regions of low densities as in the bonding regions, which should be described by the multipole model. To avoid bias, the 1/sigma^2 weights are used in most refinements against F at the expense of prominent S-values.

Hirshfeld Test: Another quality criterion is the test on the thermal motion parameters. This test is called difference of mean-square displacement amplitudes (DMSDAs), rigid-bond, or Hirshfeld test (5) [32]. The thermal parameters are tested against the rigid-body motion model [33]. If $z_{A,B}^2$denotes the mean-square displacement amplitude of atom A in the direction of atom B, for every covalently bonded pair of atoms A and B the following equation should be fulfilled:

$$\Delta_{A,B} = z_{A,B}^2 - z_{B,A}^2 \tag{5}$$

This implies that the two bonded atoms should liberate nearly equally in the bond direction. However, this rule is only strictly obeyed for atoms of the same mass. When a proper deconvolution of the ED from thermal motion is given, the DMSDAs are smaller than 1×10^{-3} Å^2 for atoms with equal masses. This value can be higher for heteronuclear bonds. If this value is significantly exceeded for homonuclei bonds, bias by unresolved valence density asphericities or an unrecognized disorder is indicated. Verification of the model and the anisotropic displacement parameters by this test strengthens the confidence in the experimentally determined ED.

3 Bond Descriptors from Charge Density Distributions

Once a charge density distribution has been obtained experimentally, various chemical and physical properties can be derived. All of these properties directly depend on the EDD.

3.1 *Static Deformation Density*

A direct inspection of the modeled density $\rho(\mathbf{r})$ itself is not very meaningful in almost all cases because the density is dominated by the core electrons and the effects of bonding are only slightly visible. Therefore, difference densities are widely applied to amplify the features of bonding. A commonly used function is the static deformation density $\Delta\rho_{\mathrm{static}}(\mathbf{r})$, which is defined as the difference between the thermally averaged density from the multipole model $\rho_{\mathrm{MM}}(\mathbf{r})$ and the spherically averaged density from the IAM $\rho_{\mathrm{IAM}}(\mathbf{r})$.

$$\Delta\rho_{\mathrm{static}}(\mathbf{r}) = \rho_{\mathrm{MM}}(\mathbf{r}) - \rho_{\mathrm{IAM}}(\mathbf{r}) \tag{6}$$

This deformation density is based on the functions and populations from the aspherical multipole refinement and therefore does not include the effect of thermal smearing.

In a deformation density map, accumulations of density are visible in bonding as well as in the lone-pair regions. This is expected, as these features are only described within the MM and not within the IAM. Thus, deformation density maps can be used to check established chemical concepts (Fig. 4).

Additionally, these maps have great diagnostic potential and are routinely used to check the quality of an analysis by a comparison of the static deformation densities from X-ray data and the theoretically derived density from the wave function.

By comparing experimental densities with those from periodic theoretical calculations, shortcomings in either method become apparent. For example, expected features cannot always be seen. Elements with more than half-filled valence shells lack bonding features in the deformation densities due to the neutral spherically averaged reference atom which contains more than one electron in each

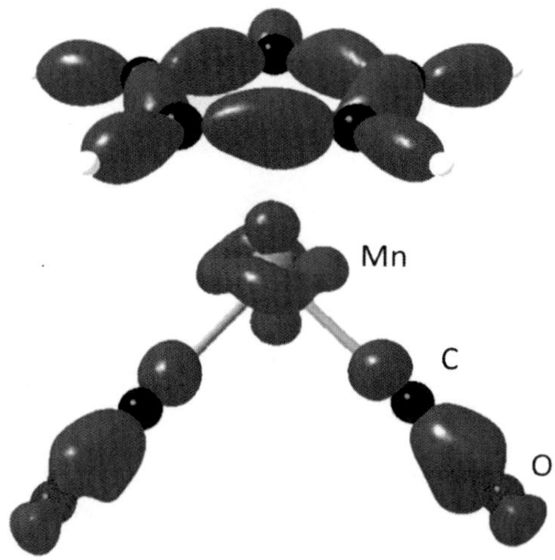

Fig. 4 Static deformation density in [{Cp(CO)$_2$Mn}$_2$(μ-B*t*Bu)] at the Mn atom, representation of a d_{z^2} density

orbital of the valence shell [34, 35]. On the contrary, in compounds with elements with less than half-filled valence shells, chemical bonding could be "amplified" by a deformation density map.

The static deformation density is, in contrast to Fourier densities, not limited to the finite resolution of the experimental data set. This leads to a high dependence on the basis set of functions applied in the refinement, and thus introduces bias. To reduce this bias, great accuracy has to be bestowed on the appropriate modeling of the reference molecule.

3.2 Electrostatic Potential

Nucleophilic and electrophilic regions in a molecule represent possible reaction sites for electrophiles or nucleophiles, respectively. As the electrostatic potential (ESP) provides information on their spatial arrangement in a molecule, its determination is of particular chemical interest.

The ESP at a given point in space is defined as the energy required for bringing a positive unit of charge from infinite distance to this point. It can be calculated independently from the crystal environment, applying the formalism of Su and Coppens [36]. As electrostatic forces are relatively long-range forces, they determine the path a reactant takes to reach the reactive site of a molecule. Hence, in chemical terms, nucleophilic reagents are attracted to regions with positive potential while electrophiles approach the negative. The direct relation between reactivity and the ESP holds well for hard Lewis acids/bases. In these reagents, the reactivity is mainly driven by electron concentrations (lone pairs) or depletions, which is directly reflected by the ESP. For soft electro- or nucleophiles, the

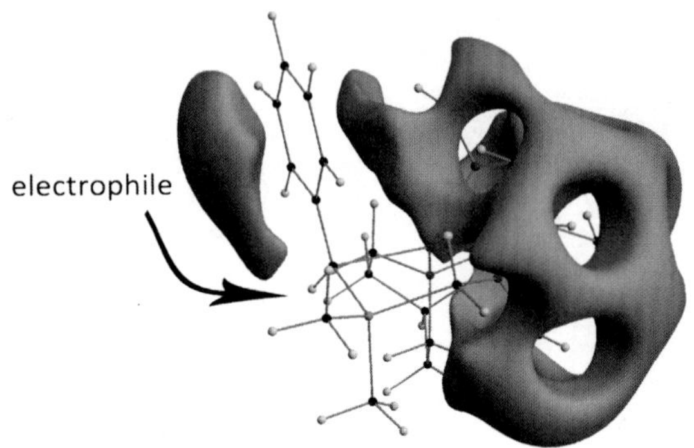

Fig. 5 Isosurface of the electrostatic potential depicted at the −0.1 e Å−1 level in a α-lithiated benzylsilane (taken from [37])

reactivity is first of all controlled by the polarizability, which may not cause direct and simple to interpret characteristics of the ESP.

In addition, one has to keep in mind that the (experimentally derived) ESP, which is calculated from a density distribution fitted by a multipole model, has to be understood as a molecular ESP in a crystal environment. This "molecular" ESP is therefore "contaminated" by the crystal field since the fitted density model which is underlying the ESP is a molecular density literally "cut" from the crystal density (Fig. 5).

3.3 d-*Orbital Populations*

The radial functions applied within the multipole refinement resemble the radial functions of orbitals. However, they are not equivalent. Therefore, the multipole populations do in general not directly correspond to the orbital populations in a given system. Nevertheless, for transition metals, it is feasible to calculate the *d*-orbital population from the least-squares multipole population coefficients [38].

This relation is based on the assumption that the overlap between metal atom and ligand orbitals is very small. As the molecular orbitals at the metal are then very similar to atomic orbitals, the *d*-orbitals can be represented by single Slater type orbitals. Therefore, the multipolar density at the transition metal atom can – to a good approximation – be calculated from the population of the outer valence shells of the atom. The relationship between *d*-orbital occupancies and multipole population parameters is derived from the equivalence of two alternative descriptions of the atomic ED.

3.4 *The Quantum Theory of Atoms in Molecules*

The objectives to extract information from the modeled density have increased to the same degree as the number of approaches to model the density. In the beginning of ED studies, the examination of the static deformation densities was the main basis of bonding type evaluation [1, 39–42]. Their results provided the first confirmation that bonding features were indeed accessible by accurate X-ray diffraction methods. However, it has been shown that the static deformation density cannot serve as the only mean of analysis [43], as the result obtained is biased by the model dependence of the promolecule describing the spherical ED [44, 45]. Nevertheless, together with the physical properties directly available from the ED, the dipole moment, and the ESP, this method already provided very interesting insights into the bonding situations of molecules.

However, the real breakthrough could only be achieved because of the advent of the topological analysis according to Bader's quantum theory of atoms in molecules (QTAIM) [46]. The assumption that all demanded information is inherent in the ED leads to the approach to closely examine the observable *density* itself. However, the distinct density concentrations at the nuclei cache the interesting – albeit very small – features in the bonding regions. A mathematical answer was found for this problem: Small features within a function can be enhanced when the derivatives are used and sketched (Fig. 6). As a consequence, it became more and more common to examine the first- and second-order derivative of the EDD topologically to obtain information on the bonding features within a molecule. Properties of the ED at certain defined points (called the critical points, CPs) within this density distribution can serve as measures for the character of bonds, while concurrently a physically meaningful separation of the molecule into its atoms allows the determination of, e.g., atomic volumes and charges.

The QTAIM method initiated a vigorous debate on chemical bonding [47–53] and provides a better acceptance of charge density studies. One of the reasons for this development certainly is the nature of the results that can be obtained via this topological analysis which can directly be applied to chemical concepts. The density, a quantum chemical property of matter, can now be interpreted in classical chemical terms. A classification of bonding in covalent or ionic, in double or single, or even more advantageous in grades thereof, becomes feasible. Second, the QTAIM provides a unique feature, its comparability. As the topological analysis can be performed on theoretically, wave function-based as well as on experimentally diffraction-based densities, the results can straightforwardly be compared.

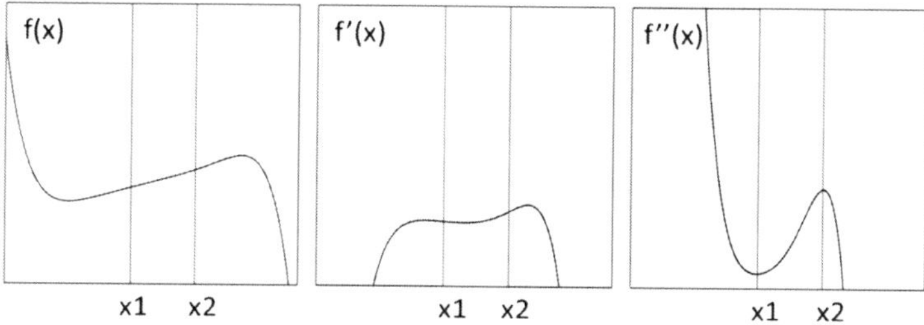

Fig. 6 Small features within a function can be amplified when inspecting its derivatives

That way, theory and experiment are of mutual benefit to examine their methods either way and identify shortcomings in either approach [54, 55]. To our opinion, it is primarily this feature that makes Bader's QTAIM the most important existing tool to examine the EDD and the resulting features of bonding that can be derived from there.

Within the framework of Bader's QTAIM, the link between the characteristics of chemical bonding and the topological properties has widely been achieved for molecules containing light atoms, i.e., those including second row elements. However, the derived correlations cannot be extended straightforwardly to organometallic compounds since bonds to transition metals display a different, less-pronounced, and much narrower spectrum of topological indices at the CPs [5].

This is due to the different valence electron configuration in transition metals compared to main group elements. Additionally, it has been shown that for very polar bonds an examination of the properties solely at the CPs is not decisive [56–60]. Therefore, new descriptors – though also based on the QTAIM – like, e.g., energy densities [61, 62] or the source function [63–68] have been introduced. Numerous descriptors are nowadays available for the examination of bonding features.

3.4.1 Topological Atoms and Atomic Charges

Bader's QTAIM is based on the assumption that the properties of a molecule can be described as the sum of the properties of its atoms. Therefore, an unambiguous definition of an atom is compulsory.

Mathematically, the density $\rho(\mathbf{r})$ of a molecule is a scalar field and its topology is best examined by an analysis of its gradient vector field. The gradient is defined as:

$$\nabla\rho(\mathbf{r}) = \mathbf{i}\frac{\partial\rho}{\partial x} + \mathbf{j}\frac{\partial\rho}{\partial y} + \mathbf{k}\frac{\partial\rho}{\partial z} \tag{7}$$

A gradient path (also called trajectory) is always perpendicular to the contour lines of $\rho(\mathbf{r})$ and follows the largest increase in $\rho(\mathbf{r})$. Therefore, it must originate from a minimum or saddle point (minimum in at least one direction) and terminate at a maximum or saddle point (maximum in at least one direction) of $\nabla\rho(\mathbf{r})$. All trajectories ending at one maximum belong to the same basin, which represents an atom in a molecule. This basin is bordered by a surface, which is not crossed by any trajectory.

$$\nabla\rho(\mathbf{r}) \cdot n(\mathbf{r}) = 0 \tag{8}$$

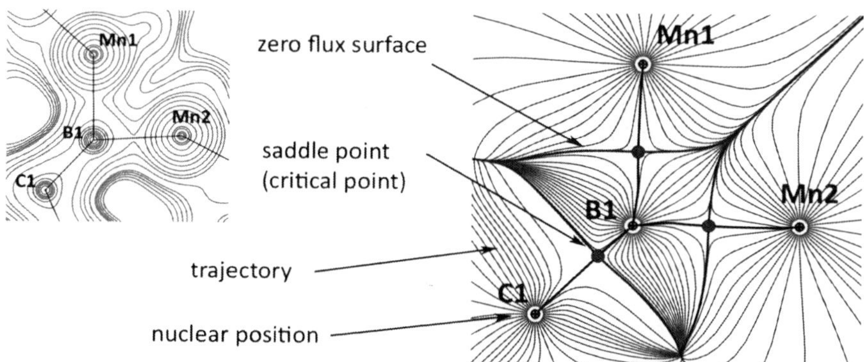

Fig. 7 ED distribution $\rho(\mathbf{r})$ in the Mn_2B plane of $[\{Cp(CO)_2Mn\}_2(\mu\text{-}BtBu)]$ (*left*) and the corresponding gradient vector field (*right*) with the trajectories (*right*) being perpendicular to the contour lines of $\rho(\mathbf{r})$ (*left*) [54]

As all points of this surface have a vanishing scalar product of $\rho(\mathbf{r}) \cdot n(\mathbf{r})$, with $n(\mathbf{r})$ being the normal vector to the surface, it is called the *zero flux surface*. It represents the border of an atomic basin and that way defines the shape of the atom (Fig. 7).

This mathematical definition of an atom allows an integration of the ED over the volume of the atomic basin. The result is the charge of the topological atom. As the volume of this topological atom depends on the bonding situation of the atom, for example on the electronegativity of the neighboring atom or the oxidation state, these charges originate from bond polarization effects as well as from charge transfer between the atomic cores. Therefore, charges following Bader's partitioning scheme are higher than those obtained by other methods [69, 70].

All determined integrated charges have to sum up to the overall charge of the total fragment (electroneutrality), which is – in the case of an uncharged molecule – zero. The electroneutrality criterion has to be fulfilled to have a reliable integration. Another quality criterion is the Lagrangian, which is indicative of the accuracy of the integrated charge [29].

3.4.2 Critical Points

The definition of the atomic basins already contained another crucial element of Bader's QTAIM. The start and end points of a gradient path are extreme values in $\rho(\mathbf{r})$. These extrema (maxima, saddle points, or minima in the ED) all have a vanishing gradient of $\rho(\mathbf{r})$. They are called critical points (CPs) and can be divided into core-, bond-, ring-, and cage-critical points (CCPs), depending on the nature of the extremum. The classification of the nature of the extremum can be performed with the help of the second-order derivative of the density. The nine second-order derivatives of $\rho(\mathbf{r})$ form the Hessian matrix:

$$H(\mathbf{r}) = \begin{pmatrix} \frac{\partial^2 \rho(\mathbf{r})}{\partial x^2} & \frac{\partial^2 \rho(\mathbf{r})}{\partial x \partial y} & \frac{\partial^2 \rho(\mathbf{r})}{\partial x \partial z} \\ \frac{\partial^2 \rho(\mathbf{r})}{\partial y \partial x} & \frac{\partial^2 \rho(\mathbf{r})}{\partial y^2} & \frac{\partial^2 \rho(\mathbf{r})}{\partial y \partial z} \\ \frac{\partial^2 \rho(\mathbf{r})}{\partial z \partial x} & \frac{\partial^2 \rho(\mathbf{r})}{\partial z \partial y} & \frac{\partial^2 \rho(\mathbf{r})}{\partial z^2} \end{pmatrix} \tag{9}$$

In its diagonalized form, the Hessian matrix provides the three eigenvalues λ_1, λ_2, λ_3 (with $\lambda_1 \leq \lambda_2 \leq \lambda_3$), which indicate the curvature of $\rho(\mathbf{r})$ along the main curvature axes at the point $\mathbf{r}$.

$$D(\mathbf{r}_{\mathrm{CP}}) = \begin{pmatrix} \frac{\partial^2 \rho(\mathbf{r})}{\partial x'^2} & 0 & 0 \\ 0 & \frac{\partial^2 \rho(\mathbf{r})}{\partial y'^2} & 0 \\ 0 & 0 & \frac{\partial^2 \rho(\mathbf{r})}{\partial z'^2} \end{pmatrix} = \begin{pmatrix} \lambda_1 & 0 & 0 \\ 0 & \lambda_2 & 0 \\ 0 & 0 & \lambda_3 \end{pmatrix} \tag{10}$$

A critical point in $\rho(\mathbf{r})$ is classified by the rank m (the number of nonzero eigenvalues λ_i) and the signature n (the algebraic sum of the signs of the eigenvalues λ_i) of the Hessian matrix. For a rank $m = 3$, there are only four possible types of CPs with (m,n) (Table 1).

Within this nomenclature, a local maximum is therefore a (3,−3) critical point. Starting from this point, the density decreases in each direction, and thus the curvature is negative in all three space directions. These points which are the endpoints of all neighboring gradient paths (see above) are called attractors and are usually associated with the nuclear position.

There are two possible kinds of saddle points within the ED distribution. The first kind of saddle point has two negative and one positive eigenvalues. These points are called (3,−1) critical points. This implies a maximum of $\rho(\mathbf{r})$ in two and a minimum in one direction, which can be found along bonds between two atoms

Table 1 Classification of critical points in $\rho(\mathbf{r})$

(m,n)	Topology in $\rho(r)$	λ_i	Interpretation	Type
(3,−3)	Local maximum	All $\lambda_I < 0$	Nuclear position	Nuclear position (NP)
(3,−1)	Maximum in two directions	Two $\lambda_I < 0$	Chemical bond	Bond critical point (BCP)
	Minimum in one direction	One $\lambda_I > 0$		
(3,+1)	Maximum in one direction	One $\lambda_I < 0$	Center of a ring of connected atoms	Ring critical point (RCP)
	Minimum in two directions	Two $\lambda_I > 0$		
(3,+3)	Local minimum	All $\lambda_I > 0$	Center of a cluster of connected atoms	Cage-critical point (CCP)

(bond critical point, BCP). The density increases from the CP toward each nuclear position but decreases in the two other directions. The gradient path following the maximum of the density from the CP to the core position is called the bond path (BP). All BPs of a molecule represent the molecular graph. In terms of the QTAIM, a BCP and its associated BP are the necessary and sufficient condition for the existence of chemical bonding [46, 71]. However, it has to be stressed out that bonding does not necessarily mean a classical Lewis two-center two-electron bond but all kinds of bonded interactions between two atoms [52]. BPs are therefore also referred to as privileged exchange channels [51].

The second possible kind of saddle point in $\rho(\mathbf{r})$ is characterized by two positive and one negative eigenvalues. Thus, these (3,+1) critical points appear when the density is minimal in two directions and decreases perpendicular to this plane. Such a scenario is often found in the center of ring systems, e.g., benzene. Here, in the center of the ring, a (3,+1) critical points is found, as the value of $\rho(\mathbf{r})$ starting from this CP is increasing in each direction of the C_6 ring and decreasing perpendicular to it. (3,+1) critical points are therefore called ring critical points (RCPs).

Local minima in $\rho(\mathbf{r})$, where all three eigenvalues are positive, always appear in the middle of a cage structure. Hence these (3,+3) critical points are called CCPs.

The reliability of the number of CPs found in a structure can be checked by the Poincaré–Hopf equation [72, 73].

$$n_{\mathrm{AP}} - n_{\mathrm{BCP}} + n_{\mathrm{RCP}} - n_{\mathrm{CCP}} = 1 \tag{11}$$

3.4.3 The Laplacian

As already pointed out before, the topology of the total ED is dominated by the contributions of the core electrons. Therefore, manifestations of paired electrons from bonding or lone pairs are difficult to detect. The amplification of small changes in the topology of the EDD is achieved via the second-order derivatives as formulated in the Hessian matrix. The Laplacian $\nabla^2\rho(\mathbf{r})$ is the trace of the Hessian matrix:

$$\nabla^2\rho(\mathbf{r}) = \frac{\partial^2\rho(\mathbf{r})}{\partial x^2} + \frac{\partial^2\rho(\mathbf{r})}{\partial y^2} + \frac{\partial^2\rho(\mathbf{r})}{\partial z^2} = \lambda_1 + \lambda_2 + \lambda_3 \tag{12}$$

The value of the Laplacian displays whether a charge concentration ($\nabla^2\rho(\mathbf{r}) < 0$) or depletion ($\nabla^2\rho(\mathbf{r}) > 0$) is present. Maxima in the negative Laplacian, (3,−3) critical points in $\nabla^2\rho(\mathbf{r})$, are therefore indicative of local charge concentrations, called valence shell charge concentrations (VSCCs). These concentrations stem from bonding electron pairs or nonbonding VSCCs (lone pairs) [74].

The spatial arrangement of the VSCCs can be used to determine the density-related bonding geometry of an atom [75–78]. Hybridization can thus much better

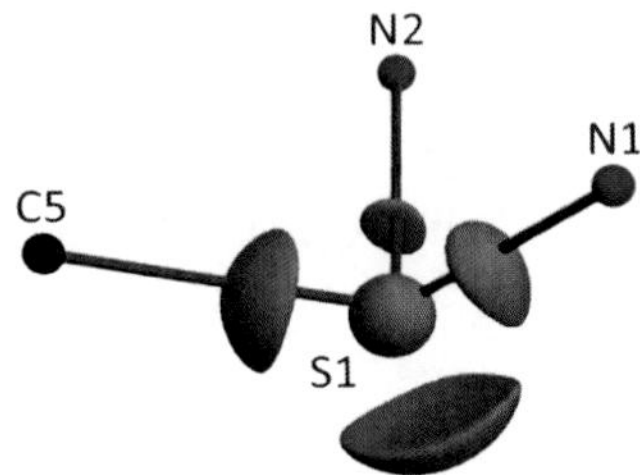

Fig. 8 Isosurface representation of the VSCCs at the sulfur atom S1 in a lithium sulfur ylide. Three of them are directed toward the bonded neighbors and one NBCC is oriented as expected for a sp^3-hybridized sulfur atom. The VSCCs around the sulfur atom include angles that range from 102.2° to 107.5°

be understood than from the traditional interatomic vectors as VSCCs stand for bonding as well as nonbonding contributions (Fig. 8). The use of the density-related bonding angles leads to a higher agreement with the ones anticipated from the VSEPR theory [79–82].

During the formation of a bond, the VSCCs of the corresponding atoms are induced to turn toward each other. Covalent bonds are characterized by an overlapping of the valence shells, more specifically the VSCCs, of the bonding partners. This causes an accumulation of charge density ($\nabla^2\rho(\mathbf{r}_{\mathrm{BCP}}) < 0$) in the bonding region and therefore at the BCP. Because of this interaction between the valence shells, covalent bonds are also called *open shell (or shared) interactions*.

The formation of ionic bonds does not induce an alignment of the VSCCs of the bonding partners. On the contrary, there is a charge depletion at the electropositive atom and a charge concentration at the more electronegative atom. The BCP is shifted toward the charge depletion at the electropositive atom ($\nabla^2\rho(\mathbf{r}_{\mathrm{BCP}}) > 0$). Graphically only one VSCC is visible, which is attributed to the electronegative atom. As interactions occur here between atoms or between ions in a closed-shell status (i.e., with almost no distortion from the atomic or ionic electronic configuration having a given electronic shell filled), they are called *closed-shell interactions*.

In the case of a covalent but very polar bond, the BCP is shifted toward the less electronegative atom. Both VSCCs are visible, but – depending on the strength of the polarization – a more or less pronounced coalescence of the VSCCs is observed.

The characterization of bonds by the sign of the Laplacian at the BCP is not always unambiguous. This is especially the case for weak bonds. The reason for this is the flat shape of the ED function along the BP. Therefore, the minimum, defined as the BCP, can only be estimated within a certain error tolerance and the BCP shows a positional uncertainty. Thus, if regarding the value of the Laplacian at this discrete point without taking the environment into account, misinterpretations are not excluded. Especially for a very polar bond, where the BCP does not lie in the middle of the BP but is shifted toward the less electronegative atom, the Laplacian can have a zero crossing close to the BCP. If the value of the Laplacian at the BCP in polar bonds is close to zero, no clear classification of the bonding is feasible. For the characterization, the Laplacian distribution along the whole BP should therefore be regarded.

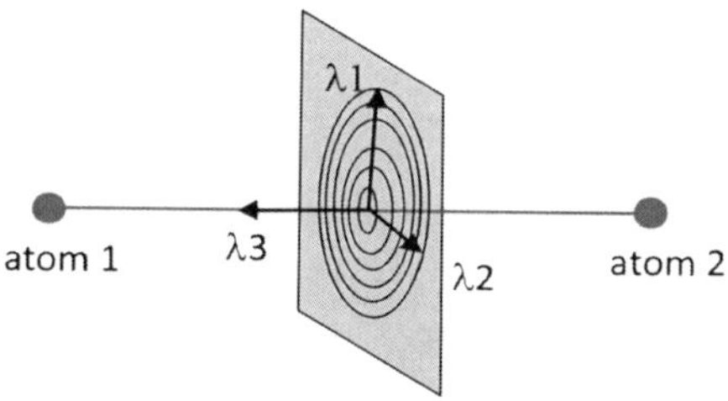

Fig. 9 Spatial orientation of the eigenvalues λ_i

3.4.4 The Ellipticity

Apart from the sum of eigenvalues, the Laplacian, the curvature of $\rho(\mathbf{r})$ can give interesting insights into the bonding situation. This curvature can be inspected by an examination of the proportion of the eigenvalues, as represented by the ellipticity:

$$\varepsilon = \frac{|\lambda_1|}{|\lambda_2|} - 1 \tag{13}$$

λ_1 and λ_2 are perpendicular to the bond axis and are the negative eigenvalues of the Hessian matrix. As λ_1 is defined to be larger or equal λ_2, ε is always positive or zero. An ideal single bond (σ-bond) has a perfect rotational cylindrical symmetry along the nuclei connecting line. λ_1 equals λ_2 and ε at the BCP is zero. Values for ε that are larger than zero indicate double bonding or deformation (Fig. 9).

For example, ε increases from ethane (0.00), to benzene (0.23), to ethene (0.45). Thus the ellipticity contains information on the bond order. For triple bonds rotational symmetry is again present and ε is close to zero.

The ellipticity of a bond can be inspected at the BCP as well as along the whole BP. For very polar bonds, however, the examination along the whole BP is compulsory to estimate the reliability of the value. Bader et al. showed that values for this property change regularly along the whole path in heteronuclear bonds [83].

3.4.5 The Source Function

Bader and Gatti showed that it is possible to regard the ED at any point within a molecule to consist of contributions from a local source operating at all other points of the space [63]. This interpretation affords a measure of the relative importance of an atom's or group's contribution to the density at any point.

For further information on this topic and the possibilities of interpreting the source function chemically we refer to the contribution by Gatti [84].

3.4.6 The Electronic Energy Density

The electronic energy density $E(\mathbf{r})$ of the ED distribution is defined as

$$E(\mathbf{r}) = G(\mathbf{r}) + V(\mathbf{r}) \tag{14}$$

In this equation, $V(\mathbf{r})$ is the potential energy density, including exchange, at the point $\mathbf{r}$ and $G(\mathbf{r})$ is a local one-electron kinetic energy density.

The electronic energy density is a straight criterion for the recognition of the atomic interaction type: $E(\mathbf{r}) < 0$ at the BCP is observed in shared-type (covalent) atomic bonding, while $E(\mathbf{r}) \geq 0$ is observed in purely closed-shell (ionic) interactions [43].

4 Conclusion

This chapter is only able to provide a small overview far from being comprehensive on some selected descriptors of the charge density distributions. Some aspects will be explained in more detail in the according chapters. In addition, there are numerous fundamental readable reviews on this topic that should hereby be advised to the dedicated reader [1, 3–9]. They provide a more detailed and if applicable more mathematical view on the topics described here. In quintessence, it still remains difficult to directly compare the heuristic concept of the well-established Lewis diagram with the physically deep-routed QTAIM BP grounded molecular graph, but clearly the latter provides a much more meaningful and detailed view of the chemical bond. It definitely opens the mind beyond the limitations of the two-electron-two-center dash between two atomic symbols.

References

1. Coppens P (1985) Coord Chem Rev 65:285–307
2. Coppens P (1997) X-Ray charge densities and chemical bonding. Oxford University Press, Oxford
3. Koritsanszky T, Coppens P (2001) Chem Rev 101:1583–1627
4. Lecomte C, Souhassou M, Pillet S (2003) J Mol Struct 647:53–64
5. Macchi P, Sironi A (2003) Coord Chem Rev 238–239:383–412
6. Coppens P (2005) Angew Chem 117:6970–6972
7. Gatti C (2005) Z Kristallogr 220:399–457
8. Stalke D, Ott H (2008) Nachrichten aus der Chemie 56:131–135
9. Gatti C, Macchi P (Eds) (2012) Modern charge density analysis. Springer, Heidelberg
10. Pauling L (1932) J Am Chem Soc 54:988–1003
11. Bragg WH, Bragg WL (1913) Proc R Soc Lond A 88:428–438
12. Hartree DR (1925) Philos Mag 50:289–306
13. Sheldrick GM (2008) Acta Crystallogr A 64:112–122

14. Kaupp M, Metz B, Stoll H (2000) Angew Chem 112:4780–4782
15. Power PP (2007) Organometallics 26:4362–4372
16. Power PP (2010) Nature 463:171–177
17. Debye P (1915) Ann Phys 351:809–823
18. Dawson B (1964) Acta Crystallogr 17:990
19. Stewart RF (1968) J Chem Phys 48:4882–4889
20. Stewart RF (1969) J Chem Phys 51:4569–4577
21. Stewart RF (1973) Chem Phys Lett 19:462–464
22. Stewart RF (1973) J Chem Phys 58:1668–1676
23. Hirshfeld FL (1971) Acta Crystallogr B 27:769–781
24. Harel M, Hirshfeld FL (1975) Acta Crystallogr A 31:162–172
25. Hansen NK, Coppens P (1978) Acta Crystallogr A34:909–921
26. Clementi E, Roetti C (1974) Atom Data Nuc Data Tab 14:177–478
27. Su Z, Coppens P (1998) Acta Crystallogr A54:646–652
28. Macchi P, Coppens P (2001) Acta Crystallogr A57:656–662
29. Volkov A, Macchi P, Farrugia LJ, Gatti C, Mallinson PR, Richter T, Koritsanszky T (2006) XD2006, a computer program package for multipole refinement, topological analysis of charge densities and evaluation of intermolecular energies from experimental or theoretical structure factors
30. International Tables for Crystallography (1996) In: Hahn T (ed) Kluwer Academic Publishers, Dordrecht, Boston, London
31. Meindl K, Henn J (2008) Acta Crystallogr A 64:404–418
32. Hirshfeld FL (1976) Acta Crystallogr A 32:239–244
33. Schoemaker V, Trueblood KN (1968) Acta Crystallogr B 24:64–77
34. Savariault JM, Lehmann MS (1980) J Am Chem Soc 102:1298
35. Dunitz JD, Schweizer WB, Seiler P (1983) J Am Chem Soc 105:7056
36. Su Z, Coppens P (1998) Acta Crystallogr A 54:646–652
37. Ott H, Däschlein C, Leusser D, Schildbach D, Seibel T, Stalke D, Strohmann C (2008) J Am Chem Soc 190:11901–11911
38. Holladay A, Leung P, Coppens P (1983) Acta Crystallogr A39:377–387
39. Eisenstein M (1979) Acta Crystallogr B 35:2614–2625
40. Hirshfeld FL, Hope H (1980) Acta Crystallogr B 36:406–415
41. Kirfel A, Gupta A, Will G (1980) Acta Crystallogr B 36:1311–1319
42. Kirfel A, Will G (1981) Acta Crystallogr B 37:525–532
43. Cremer D, Kraka E (1984) Angew Chem 96:612–614
44. Coppens P, Stevens ED (1977) Adv Quantum Chem 10:1–35
45. Coppens P (1977) Angew Chem 84:33–41
46. Bader RFW (1990) Atoms in molecules - a quantum theory. Oxford University Press, New York
47. Shu-Guang W, Schwarz WHE (2000) Angew Chem 112:1827–1832
48. Matta CF, Hernández-Trujillo J, Tang T-H, Bader RFW (2003) Chem Eur J 9:1940–1951
49. Haaland A, Shorokhov DJ, Tverdova NV (2004) Chem Eur J 10:4416
50. Henn J, Leusser D, Stalke D (2007) J Comput Chem 28:2317–2324
51. Pendás AM, Francisco E, Blanco MA, Gatti C (2007) Chem Eur J 13:9362–9371
52. Bader RFW (2009) J Phys Chem A 113:10391–10396
53. Jacobsen H (2009) J Comput Chem 30:1093–1102
54. Flierler U, Burzler M, Leusser D, Henn J, Ott H, Braunschweig H, Stalke D (2008) Angew Chem 120:4393–4397
55. Götz K, Kaupp M, Braunschweig H, Stalke D (2009) Chem Eur J 15:623–632
56. Kocher N (2003) Experimental charge density studies of highly polar bonds. PhD thesis, Würzburg, Germany
57. Kost D, Gostevskii B, Kocher N, Stalke D, Kalikhman I (2003) Angew Chem 115:1053–1056

58. Kocher N, Henn J, Gostevskii B, Kost D, Kalikhman I, Engels B, Stalke D (2004) J Am Chem Soc 126:5563–5568
59. Kocher N, Leusser D, Murso A, Stalke D (2004) Chem Eur J 10:3622–3631
60. Kocher N, Selinka C, Leusser D, Kost D, Kahlikhman I, Stalke D (2004) Z Anorg Allg Chem 630:1777–1793
61. Abramov YA (1997) Acta Crystallogr A 53:264–272
62. Stash A, Tsirelson V (2002) J Appl Crystallogr 35:371–373
63. Bader RFW, Gatti C (1998) Chem Phys Lett 287:233–238
64. Overgaard J, Schiøtt B, Larsen FK, Iversen BB (2001) Chem Eur J 7:3756–3767
65. Gatti C, Cargnoni F, Bertini L (2003) J Comput Chem 24:422–436
66. Gatti C, Bertini L (2004) Acta Crystallogr A60:438–449
67. Gatti C, Lasi D (2007) Faraday Discuss 135:55–78
68. Sørensen J, Clausen HF, Poulsen RD, Overgaard J, Schiøtt B (2007) J Phys Chem A 111:345–351
69. Wiberg KB, Breneman CM (1990) J Am Chem Soc 112:8765–8775
70. Messerschmidt M, Wagner A, Wong MW, Luger P (2002) J Am Chem Soc 124:732
71. Bader RFW (1998) J Phys Chem A 102:7314–7323
72. Hopf H (1927) Math Ann 96:209–224
73. Collard K, Hall GG (1977) Int J Quantum Chem 12:623–637
74. Tang T-H, Bader RFW, MacDougall PJ (1985) Inorg Chem 24:2047–2053
75. Gillespie RJ (1972) Molecular geometry. Reinhold Van Nostrand, London
76. Gillespie RJ, Hargittai I (1991) The VSEPR model of molecular geometry. Allyn and Bacon, Boston, London
77. Bytheway I, Gillespie RJ, Tang T-H, Bader RFW (1995) Inorg Chem 34:2407–2414
78. Gillespie RJ, Robinson EA (1996) Angew Chem 108:539–560
79. Bader RFW, Gillespie RJ, MacDougall PJ (1989) Mol Struct Energetics 11:1–51
80. MacDougall PJ, Hall MB, Bader RFW, Cheeseman JR (1989) Can J Chem 67:1842–1846
81. Bader RFW, Johnson S, Tang T-H, Popelier PLA (1996) J Phys Chem 100:15398–15415
82. Leung W-P, Wang Z-X, Li H-W, Yang Q-C, Mak TCW (2001) J Am Chem Soc 123:8123–8124
83. Cheeseman JR, Carroll MT, Bader RFW (1988) Chem Phys Lett 143:451
84. Gatti C (2010) The source function descriptor as a tool to extract chemical information from theoretical and experimental electron densitie. Struct Bond DOI: 10.1007/430_2010_31

Struct Bond (2012) 146: 21–52
DOI: 10.1007/430_2011_70
© Springer-Verlag Berlin Heidelberg 2012
Published online: 27 March 2012

Modeling and Analysis of Hydrogen Atoms

Anders Østergaard Madsen

Abstract Hydrogen atoms are elusive seen from the point of view of the X-ray crystallographer. But they are also extremely important, being involved in a wealth of intermolecular interactions and thereby defining the way molecules interact. Most experimental charge density studies are performed on compounds containing hydrogen, yet a commonly accepted strategy to deal with these elusive but so important atoms is only just about to surface. We review the efforts to determine a strategy for the modeling of hydrogen atoms, as well as a number of recent studies where the modeling of hydrogen atoms has had a major impact on the chemical conclusions drawn from analysis of the experimental charge densities.

Keywords Charge density analysis · Debye–Waller factors · Hydrogen atoms

Contents

A. Østergaard Madsen (✉)
Department of Chemistry, University of Copenhagen, Universitetsparken 5, DK-2100 Copenhagen Ø., Denmark
e-mail: madsen@chem.ku.dk

Abbreviations

ADP Anisotropic displacement parameter
BCP Bond critical point
DFT Density functional theory
HF Hartree–Fock
IAM Independent atom model
MSD Mean square displacement
PES Potential energy surface
SDS Scattering factor for hydrogen by Stewart, Davidson, and Simpson

1 Introduction

The properties of hydrogen atoms have attracted crystallographers for many years, because of the importance of hydrogen bonding in virtually all areas of organic chemistry and structural biology. Crystallography has played – and continues to play – a major role in defining and elucidating the properties of hydrogen bonding [1].

However, because X-rays interact with the electron density in the crystal, the information about the H atom nuclear position and motion that can be derived from X-ray diffraction analysis is limited; the lack of core-electrons and the diffuse character of the valence shell make the scattering power of hydrogen poor compared to other elements. For this reason, the majority of structural studies of hydrogen bonding in molecular crystals has relied on neutron diffraction, where the scattering length of hydrogen has a magnitude comparable to the heavier elements.

In experimental charge density studies, the combined X-ray and neutron diffraction study has proven to be an important tool to study the charge density of hydrogen atoms. Unfortunately, it is not always possible to perform neutron diffraction experiments; it might be impossible to grow crystals of a sufficient size (larger than 1 mm^3), or the researchers' access to a neutron source is limited. Because the speed and quality of the X-ray diffraction experiments have increased tremendously, conducting neutron diffraction experiments have become a bottleneck which may prohibit comparative studies of charge densities in a series of related compounds.

For these reasons, many researchers have avoided the use of neutron diffraction data and limited their model of the hydrogen atom nuclear parameters to an isotropic displacement parameter, combined with a very limited description of the charge density of the hydrogen atoms. In fact, more than 80% of the studies in the period from 1999 to 2007 used this approach [2]. In this chapter, we review the efforts that has been done to elucidate the errors introduced in the charge density models by using a less than adequate description of the hydrogen atoms, as well as a number of methods to obtain a better model of hydrogen atoms, in lieu of neutron diffraction data.

This review is divided into three parts:

1. In Sect. 3, we review the work that demonstrate how an incorrect description of the nuclear parameters of H atoms leads to erroneous models of the static charge density in the crystal.
2. In Sect. 4, we review the efforts to determine the positions and thermal motion of hydrogen atoms for the use in experimental charge-density studies. We briefly describe the pros and cons of using information from neutron diffraction experiments, but focus mainly on presenting methods of estimating hydrogen anisotropic displacement parameters (ADPs) based on the analyses of the thermal motion of the non-hydrogen atoms combined with information about the internal motion of the hydrogen atoms. It is not the purpose of this chapter to compare the accuracy of these different approaches to estimate hydrogen ADPs – that has been done quite thoroughly by Munshi et al. [2].
3. In Sect. 5, we describe recent charge-density studies where the use of accurate information about the H atom position and thermal motion has been crucial, and consider the strengths and weaknesses of the different approaches to estimate the positions and thermal motion of hydrogen atoms.

2 Bias in Static Charge-Density Models Due to Incorrect Deconvolution of Thermal Motion and Incorrect Nuclear Positions of Hydrogen Atoms

Even at the lowest temperatures atoms in crystals are vibrating. The diffracted X-ray or neutron intensities correspond to the diffraction signal of the averaged electron and nuclear density $\langle \rho \rangle$. The brackets denote a double average over the displacement of atoms from their mean positions; a time average over the atomic vibrations in each unit cell, as well as a space average over all unit cells. The structure factor of a reflection **h** is given by the Fourier transform of the average density of scattering matter.

$$F(\mathbf{h}) = \int \langle \rho(\mathbf{r}) \rangle \; \exp(2\pi i \mathbf{h} \cdot \mathbf{r}) d\mathbf{r}. \tag{1}$$

The standard independent atom model (IAM) used in structural studies, where the atomic scattering factors are spherically symmetric, as well as the multipole model used in charge-density studies is constructed in order to deconvolute the thermal motion from a model of the static electron density. In this approach, the *mean thermal* electron density of an atom is considered to be the convolution of a static density $\rho_k(\mathbf{r})$ with the probability density function (PDF) p_k describing the probability of having atom k displaced from its reference position.

$$\langle\rho_k(\mathbf{r})\rangle = \int \rho_k(\mathbf{r}-\mathbf{r}_k)p_k(\mathbf{r}_k)d\mathbf{r}_k = \rho_k(\mathbf{r}) \times p_k. \tag{2}$$

The structure factors can then be approximated by a sum over contributions from all N atoms of the unit cell:

$$F(\mathbf{h}) \approx \sum_{k=1}^{N} n_k f_k(\mathbf{h})T_k(\mathbf{h})\ \exp(2\pi i\mathbf{h}\cdot\mathbf{r}_{k0}), \tag{3}$$

where f_k is the form factor of atom k and $T_k(\mathbf{h})$ is the Fourier transform of the PDF, p_k, describing the displacement of atom k from its reference position $\mathbf{r}_{k0}$. The atomic form factors are described by the multipole model. The PDF p_k and its Fourier transform $T_k(\mathbf{h})$ – the atomic Debye–Waller factor – are normally described by trivariate Gaussian functions, because it can be shown that the PDF is a Gaussian distribution if the vibrations of the atoms are harmonic [3]. A common way to express the trivariate Gaussian atomic Debye–Waller factor is in the coordinate system of the crystallographic unit cell with axes a^i, $i \in 1,2,3$:

$$T_k(\mathbf{h}) = \exp\left(-2\pi^2\sum_{i=1}^{3}\sum_{j=1}^{3} h_i a^i U^{ij} a^j h_j\right), \tag{4}$$

with

$$U^{ij} = \langle\Delta\xi^i\,\Delta\xi^j\rangle, \tag{5}$$

where $\Delta\xi^i$ are the components of the displacement vector $\mathbf{u}$ of atom k from its mean position. The elements U^{ij} of the 3×3 tensor $\mathbf{U}$ are the ADPs, and in this form the elements have dimension (length^2) and can be directly associated with the mean-square displacements of the atom considered in the corresponding directions. There exist a confusing range of different representations and nomenclature of the ADPs – an overview and some recommendations concerning the nomenclature are given by Trueblood et al. [4].

A major goal of experimental charge-density analysis is to ensure that a proper deconvolution of vibration effects and static charge density is obtained. The entire model of $\langle\rho(\mathbf{r})\rangle$ consisting of both static and dynamic elements is refined against one set of structure factors, cf. (3). Because the parameters of the multipole model correlate with the parameters describing the positions and mean square displacements of the atoms it is evident that there can be no proper description of the static charge density without a proper description of the positions and molecular motion of the atoms, and vice versa.

One of the advantages of the multipole model is that it provides a much better deconvolution of the thermal motion of non-hydrogen atoms than the IAM model. However, the situation is more complicated for hydrogen atoms. Historically, the scattering factor of hydrogen was based on the ab initio wavefunction of an isolated hydrogen atom. The electron density of an isolated H atom is much different from a bonded hydrogen atom, and the scattering factor lead to meaningless bond distances and thermal parameters for hydrogen atoms. The so-called SDS scattering factor based on the ab initio electron density of the hydrogen molecule [5] lead to a considerable improvement, and this scattering factor is today the standard in popular structure refinement programs. Even with the SDS scattering factor, covalent bond lengths to hydrogen obtained by refinement against X-ray diffraction data are roughly 20% shorter than the values obtained from refinement against neutron diffraction data. And while the SDS scattering factor allows refinement of isotropic displacement parameters, it is not possible to obtain ADPs that are in agreement with neutron diffraction data, or even physically meaningful.

This review has a special focus on the modeling of the nuclear parameters of the hydrogen atoms, however, a few comments on the modeling of the electron densities are important. Model studies of diatomic hydrides by Chandler and Spackman [6] indicated that in order to model the electron densities of the hydrogen atoms, it is important to require multipole expansions to at least the quadrupole level, with one single-exponential radial function per multipole and all exponents optimized. Numerous experimental studies including a range of the studies reviewed in this chapter indicate that the quadrupole terms are important. The radial functions are usually of the single exponential form, $r^n \exp(-\alpha r)$, with $n = 0, 1, 2$ for the monopole, dipole, and quadrupole terms, respectively. However, optimization of exponents of the radial functions is not conventional, and if performed only with a common exponent α for all the radial functions on all H atoms. This limited refinement of exponents is dictated by the quality and resolution of the experimental data.

Experimental charge densities can also be studied using maximum entropy methods. Some recent developments are described by Netzel et al. [7, 8]. The electron densities obtained by this method are mean thermal charge densities, i.e., there is no attempt to deconvolute dynamic effects from a static electron density model, and thereby one of the major challenges of the multipole model is avoided. The MEM method and the studies of thermally averaged charge densities are very interesting but out of the scope of the present review that deals with the deconvolution of the thermal motion of hydrogen atoms from the thermally averaged charge density.

2.1 Hydrogen Atom Positions

As mentioned previously, the standard SDS scattering factor [5] improved the description of the hydrogen atoms tremendously as compared to the scattering factor of an isolated H atom, however it leads to bond lengths that are about 20% too short, as compared to the values obtained from neutron diffraction data. In fact, the SDS scattering factor was designed to obtain the best possible least-squares fit against the diffraction data, and not to obtain the correct position of the hydrogen atom. Since the electron density of a bound hydrogen atom is polarized toward the bonding atom, it is difficult to obtain a better estimate of the hydrogen position with a spherically symmetric description of the electron density of hydrogen. Some efforts to provide alternative spherical scattering factors for hydrogen can be found in the literature, but they have never been thoroughly tested by the crystallographic community (e.g., [9]). Unfortunately, and perhaps contrary to expectation, the situation does not improve by refining the hydrogen positions in a conventional multipole model. A number of approaches to obtain approximate hydrogen positions for the use in charge-density studies have been proposed, and will be discussed in Sect. 4.2.

2.2 Hydrogen Atom ADPs

There is no doubt that H atoms in molecular crystals vibrate anisotropically; the amplitudes of bond-bending motion are much larger than the amplitudes of bond-stretching motion. Translational and librational molecular motion can also enhance the anisotropy. To use isotropic displacement parameters is a severe approximation, because the isotropic displacement parameters correlate with the monopole parameters of the multipole model, and the quadrupole parameters will include dynamic effects because the quadrupole deformations and the ADPs have the same symmetry. An investigation of the consequences of applying isotropic displacement parameters was given by Madsen et al. [10]; a number of molecular test systems – xylitol [11], urea [12], methylammonium hydrogen succinate monohydrate [13], and methylammonium hydrogen maleate [14]) for which highly accurate X-ray and neutron-diffraction data were available were investigated, and models including isotropic hydrogen displacement parameters were compared to reference models including ADPs derived from the neutron-diffraction experiments. A topological analysis of the static charge density models showed large differences in the electron density and curvature of intramolecular bond-critical points, not only for bonds involving hydrogen atoms. Figure 1 shows the results obtained for xylitol [11]. The bond-critical points mark the boundary between the atomic basins, defined by their zero-flux surfaces within the quantum theory of atoms in molecules (QTAIM) of Bader and coworkers [16]. The volumes and charges of the atomic basins are therefore also affected by the erroneous treatment of hydrogen atoms, which is

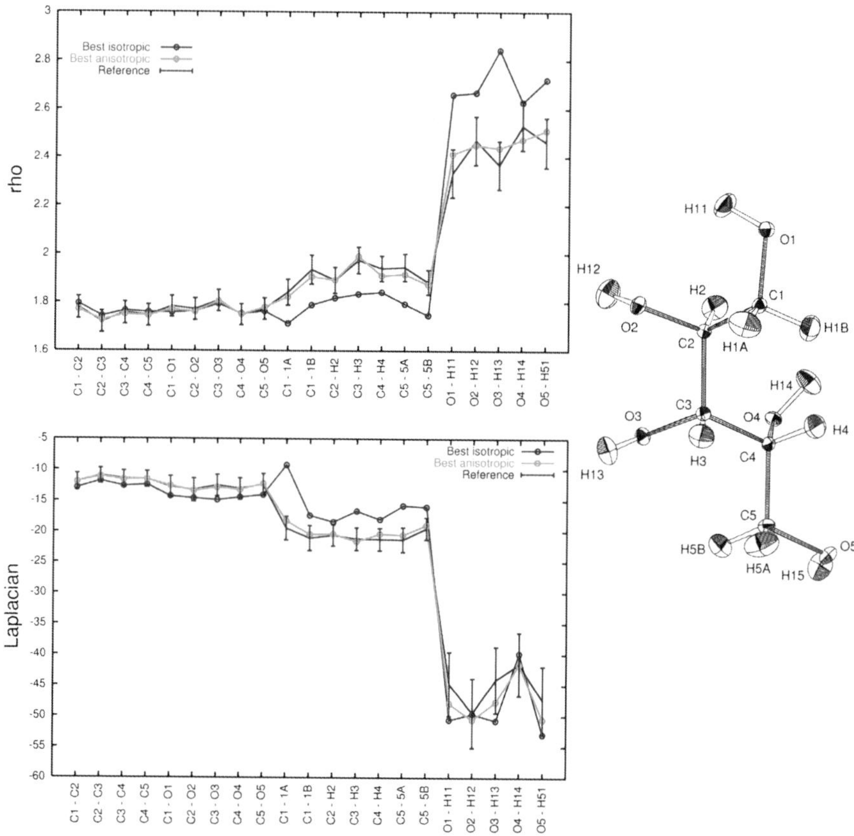

Fig. 1 Electron density [e Å^{-3}] and Laplacian [e Å^{-5}] in the intramolecular bond critical points of three xylitol models. In the reference model (*red color*) the hydrogen ADPs are based on neutron diffraction data. The error bars correspond to three s.u.s of the properties of this reference model. In the isotropic model (*blue*) the best possible isotropic description of hydrogen ADPs was employed. The *green curve* shows the bond critical points from the model using estimated H ADPs (SHADE approach). Modified version of Fig. 4 from Madsen et al. [10]

evident in the study by Hoser et al. [17]. Their work also indicates that it is only meaningful to compare topological properties between different systems when the multipole expansion of the hydrogen atoms are contracted at the same level for all systems. Similar conclusions about the modeling of hydrogen atoms were drawn by Roversi and Destro [18] for topologies and derived electrostatic properties of the electron density; when the description of the hydrogen atoms is limited to an isotropic displacement parameter combined with a contraction of the multipole expansion before the quadrupole functions, the modeling of hydrogen atoms becomes inadequate: A map of the Laplacian of the electron density of the fungal metabolite citrinin [19] shown in Fig. 2a is evidently affected by the isotropic displacement parameters and lack of quadrupole components, as compared to the more elaborate model in Fig. 2b. In the latter case, there is a clear concentration of charge (i.e., an increase in the negative region of r2q) toward the acceptor O atoms.

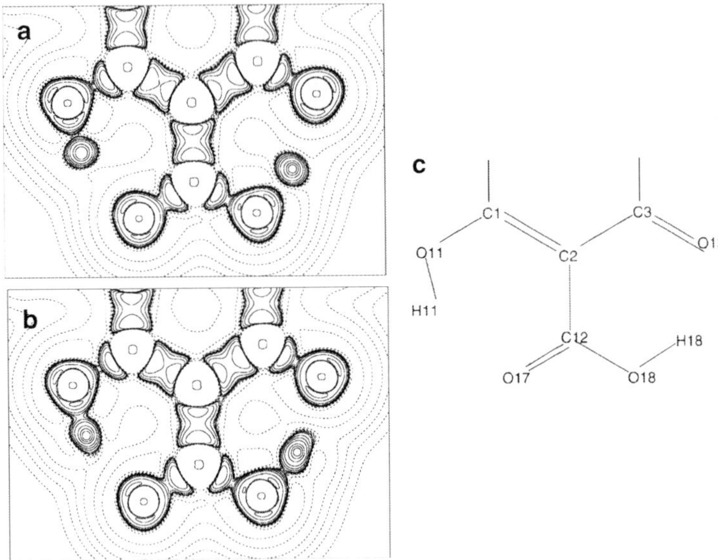

Fig. 2 Laplacian map ($-\nabla^2\rho$) for a fragment of the citrinin molecule [19], adapted from [18]. (**a**) After refinement of a model with U^{iso} and no quadrupole functions for the H atoms; (**b**) after refinement of a model with the U^{ij} and quadrupole functions added to the H atoms; (**c**) atomic numbering scheme

Roversi and Destro further demonstrate that the use of ADPs on H atoms yields electric field gradients (EFGs) at the H nuclei of L-alanine in quantitative agreement with nuclear quadrupole resonance results. Similar agreement with NQR measurements were obtained by Buergi et al. [20] in a study of benzene, as further described in Sect. 5.

The work of Mata et al. [21] examined the influence of including quadrupole functions in the description of the H atom density on the properties of intermolecular interactions. They conclude that these functions in conjunction with hydrogen ADPs have a significant impact on the interaction density, as further discussed in Sect. 5.3

To conclude, there seems to be little doubt that the thermal motion and static charge densities of hydrogen atoms are some of the most difficult to model in charge-density analysis. A proper anisotropic description of the thermal motion combined with a multipole expansion including quadrupole components seems to be the smallest adequate model.

2.3 A Comment on Using the Residual Density and R-Values to Judge the Quality of Models for Hydrogen Atoms

Inspection of residual density plots and R-values are important tools during the construction of a charge-density model. However, these tools cannot stand alone.

The least-squares refinement does not guarantee that the model is physically sound, it merely ensures that we have reached a position in parameter space where the residual is minimized (locally). In the comparison of models with and without estimated ADPs for hydrogen atoms performed by Madsen et al. [10] and Roversi and Destro [18], the change in R-values between the different models was negligible. The less advanced models, using only an isotropic displacement parameter, and in some cases only one bond-directed dipole component and no quadrupole components in the hydrogen multipole expansion are flexible enough to minimize the residual. However, because the model is not physically acceptable, important features of the charge density model are left out. An example that may make the point clear is to consider the position of the hydrogen atom in a standard IAM refinement: The position refines to give an X–H bond length which is way shorter than the position obtained using neutron diffraction data. However, constraining the model to a physically reasonable hydrogen position will probably increase the R-value and give features of negative and positive density in the residual density maps close to hydrogen atoms. Obviously, such features should be removed by improving the quality of the electron density model (e.g., using a multipole description), not by refining the position of the hydrogen atom.

3 How to Obtain H Atom Positions and Anisotropic Displacement Parameters

3.1 Information from Neutron Diffraction

A complementary neutron diffraction study is undoubtedly the most reliable source of hydrogen nuclear parameters. However neutron sources are sparse, and the low neutron flux requires millimeter-sized crystals, much larger than the ones needed for the X-ray diffraction experiments. A new generation of neutron spallation sources such as SNS at Oak Ridge National Laboratory (USA) and the ESS (Lund, Sweden) undoubtedly remove some of these barriers, i.e., by allowing much smaller samples and shorter collection times.

The different scattering phenomena and experimental conditions for the X-ray and neutron diffraction experiments cause different systematic errors: thermal diffuse scattering (TDS), extinction and absorption effects are different. The X-ray diffraction experiment is also prone to spectral truncation effects [22]. While absorption effects can be handled by a proper correction based on the morphology of the crystal, and extinction can be taken into account as part of the crystallographic model [23, 24], there is presently few options to deal with the TDS but to cool the crystals to very low temperatures, preferably using helium, to reach temperatures below 20 K [25] thereby diminishing the TDS.

The systematic differences between ADPs from X-ray and neutron diffraction experiments are discussed in detail in the literature [26, 27], and Blessing [26] provides a practical scaling procedure to account for systematic differences between the non-hydrogen ADPs derived from neutron and X-ray diffraction experiments, in order to use the coordinates and ADPs from the neutron-diffraction experiments as fixed parameters in the refinement of a charge-density model against the X-ray diffraction structure factors.

In some cases neutron diffraction data have a quality and a resolution that merits the refinement of a model including anharmonic motion. The most commonly used model is inclusion of Gram-Charlier coefficients [4, 28]. However, as noted by Kuhs [29], the refinement of these third-order cumulants requires a very extensive dataset, especially at elevated temperatures, where anharmonic motion can be considerable.

Whereas there might be systematic errors affecting the reliability of the ADPs obtained from neutron diffraction, this is much less so for the positions, which are often transferred without any corrections from the model refined against neutron data, to be used as fixed parameters in the refinement against X-ray data. In the absence of results from a neutron diffraction experiment, a number of approaches have been proposed, as discussed below.

3.2 *Estimated Hydrogen Positions*

3.2.1 High-Order Refinement

It is well known that the accuracy of the atomic positions of non-hydrogen atoms obtained from X-ray diffraction experiments can be improved by refining an IAM against the high-order reflections (i.e., $\sin(\theta)/\lambda > 0.7$ Å^{-1}, see the work by Ruysink and Vos [30]). The reason is that the relative contribution of the core electrons to the reflection intensities increases with resolution, and in this way the shifts in atomic positions due to the modeling of bonding density [31] can be diminished. However, hydrogen atoms have no core electrons, and the advantages of this approach seem less obvious for the determination of hydrogen atomic positions. Hope and Ottersen [32] tested this approach in a study of s-diformohydrazide and found considerable improvements in the bond-lengths involving hydrogen atoms. Almlöf and Ottersen [33] have given a theoretical analysis of the high-order refinement of hydrogen positions. Some improvement was also seen by Madsen et al. [10] for xylitol, however in that case other methods proved to be in closer agreement with neutron diffraction results, as described below.

3.2.2 A Polarized Hydrogen Atom

Realizing that the refinement of a spherical scattering factor against X-ray structure factors is not sufficient to provide hydrogen positions in agreement with neutron diffraction results Stewart and coworkers [34] proposed that the use of a fixed multipole expansion for a bonded hydrogen atom could be used to obtain a more accurate determination of the time proton positions in molecular crystals.

This idea has been implemented in the VALRAY program [35] and used for X-ray charge-density analysis. This *polarized hydrogen atom* consists of spherical component (single-exponential type) and a bond-directed dipole. The population of the dipole and the spherical components is kept fixed during the refinement, and only the position and isotropic displacement parameter of the hydrogen atom is refined.

Destro and coworkers have used this approach in a range of studies (e.g., [19, 36–38]) and were able to obtain X–H bond-lengths close to the typical values found in neutron-diffraction studies, although no direct comparison with neutron diffraction measurements was made.

Madsen et al. [10] tested the polarized hydrogen atom in refinements against X-ray structure factors of xylitol and compared the results to the positions obtained from neutron diffraction [11]: The polarized hydrogen atom was an improvement compared to the SDS scattering factor, but not as efficient as the posteriori elongation of X–H bond lengths to match mean values from neutron diffraction experiments.

3.2.3 Neutron Mean Values

While the high-order refinements and polarized hydrogen atom model are attempts to obtain the hydrogen positions based on the X-ray diffraction data, a more pragmatic procedure is to base the positions on statistical material from a large pool of structures determined from neutron diffraction data.

The International Tables for Crystallography [39] contains a wealth of information on X–H bond lengths based on data from the CSD database. Once the direction of the X–H bond has been established by refinement against the X-ray data, the bond can be extended to match the mean values derived from neutron diffraction experiments. This procedure has been used in numerous charge-density studies (some very recent examples are Overgaard et al. [40]) and seems to give the best possible estimate of the hydrogen positions. Madsen et al. [10] found for xylitol that the mean deviation from the neutron result was 0.012(8) Å in bond lengths, corresponding to a mean deviation in positions of 0.041(19) Å.

The X–H bond-lengths depend on the number and strength of the hydrogen bonds that the X–H group is involved in [41], and this information could in principle be used to improve the estimates of hydrogen positions.

3.2.4 A Combined Approach

Hoser et al. [17] analyzed models for a series of 1,8-bis-(dimethylamino)naphthalene (DMAN) salts of organic counter-ions. They conclude that a combined approach involving high-order refinement of the non-hydrogen atoms combined with low-order refinement of the hydrogen atoms and subsequent elongation of the X–H bond lengths to match mean-values derived from neutron diffraction studies gives the best results as compared to complementary neutron-diffraction studies.

3.3 Ab Initio/Hirshfeld Atom Refinement

A sophisticated extension of the improved aspherical scattering factors for hydrogen atoms used in the polarized H atom model is the refinement strategy of Jayatilaka and Dittrich [42], which is a least-squares refinement against X-ray diffraction data of transferable atomic densities defined in terms of "Hirshfeld atoms." The atoms are defined by using Hirshfeld's stockholder partitioning [43] of an electron density obtained from quantum mechanical calculations. The strategy was tested by refining against X-ray data for urea and benzene, and benchmarked against neutron diffraction results. The C–H and N–H bond distances are remarkably within 0.01 Å of the neutron diffraction results. The approach yields ADPs of the carbon and hydrogen atoms of benzene in excellent agreement with the neutron diffraction data. However, for urea the results are a bit ambiguous with some ADPs in excellent agreement with neutron diffraction results, others deviating more than 50%, and a large dependence on whether the applied electron density was obtained using Hartree–Fock (HF) or density functional theory (DFT) methods. This approach seems very promising, but it has to be further validated on more structures before any conclusions can be drawn as to its general applicability to obtain positions and ADPs for hydrogen atoms.

3.4 Estimated H ADPS Based on a Combination of Rigid Body Motion and Internal Motion

The method of estimating hydrogen ADPs based on a combination of rigid body motion derived from the ADPs of the non-hydrogen atoms and internal modes taken from other experiments was pioneered by Hirshfeld and coworkers [44–47]. It is based on the idea that it is possible to consider the atomic motion in molecular crystals as a combination of external rigid-body motion and internal motion corresponding to high-frequency molecular vibrations, such as bond-bending and bond-stretching modes. The analysis is based on the assumption that the rigid-body and internal motions are uncorrelated, and that the components of B, the atomic mean square displacement matrix – and thereby the U matrix of ADPs – can be obtained as a sum of the two contributions

$$U^{ij} = U^{ij}_{\text{rigid}} + U^{ij}_{\text{internal}}. \tag{6}$$

The external motion is taken from a rigid-body or segmented rigid-body analysis of the non-hydrogen ADPs, while the internal motion can be estimated from spectroscopic experiments, analysis of neutron diffraction data, or based on ab initio calculations. These three different approaches are discussed below – a thorough comparison has been given previously [2].

3.4.1 The Rigid Body Motion

Because the intermolecular forces are weak compared to intramolecular forces, the external molecular vibrations have larger amplitudes and lower frequencies than the internal modes. The atomic motion of the non-hydrogen atoms is therefore mostly due to the external molecular vibrations. A commonly used model is to regard the molecules in the crystal as vibrating as rigid bodies, independently of the surrounding molecules, and experiencing the mean potential of these molecules. This rigid-body model, used to analyze the ADPs of the non-hydrogen atoms, was introduced by Cruickshank [48] and refined by Schomaker and Trueblood [49] into what is known as the TLS model and has been used in numerous studies [50]. An excellent introduction to the TLS analysis is given by Dunitz [51].

The rigid body assumption can be relaxed in different ways to include torsional vibrations or other large-amplitude vibrations [52]. The TLS model is implemented in the THMA program of Schomaker and Trueblood, as well as in the Platon program [53]. A related normal-mode analysis is provided by the program EKRT by Craven and coworkers [54, 55], and by the program NKA of Buergi and coworkers [56]. The latter program is furthermore able to refine a model against ADPs from multi-temperature experiments [57, 58].

The quality of the rigid body model is normally judged by computing a residual defined as

$$R_w(U^{ij}) = \sum_{i,j} w_{ij} \frac{U^{ij}_{\text{measured}} - U^{ij}_{\text{TLS model}}}{U^{ij}_{\text{measured}}}, \tag{7}$$

where w_{ij} is the weight used in the least squares fit of the TLS model against the observed ADPs U^{ij}_{measured}. The TLS analysis of well-determined ADPs of truly rigid bodies (e.g., benzene) often gives $R_w(U^{ij})$ values of about 5%, especially for low-temperature studies. For less rigid systems values of 8–12% are common. For the purpose of estimating hydrogen ADPs, the TLS models of these less rigid systems still seem to give good results. Apart from this residual, an excellent way to judge the rigidity of a system is to compare the differences $\Delta_{AB}(\langle u^2\rangle)$ in mean square displacements along all atom–atom vectors in the molecule [59]. In a truly rigid body these differences should be exactly zero. However, even in rigid molecules, atoms of different mass will have small differences in internal mean square displacements due to the internal vibrations, giving rise to $\Delta_{AB}(\langle u^2\rangle)$ values on the order of 10^{-4} Å^2.

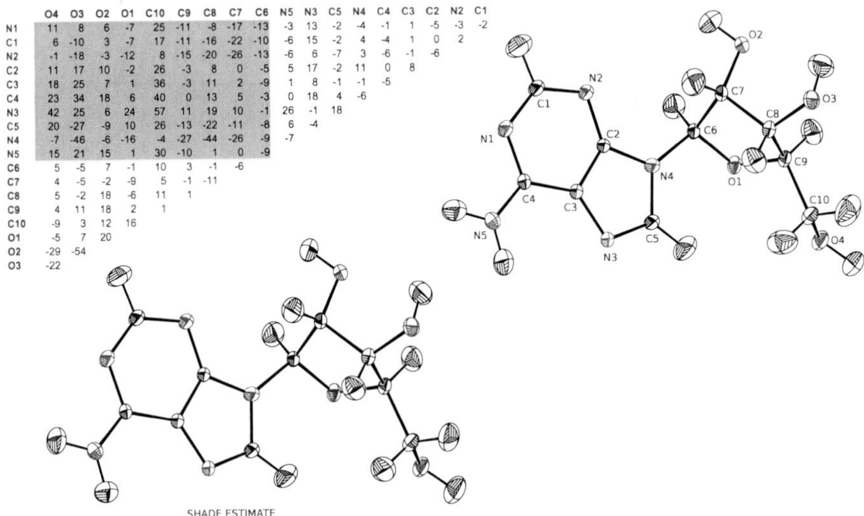

	O4	O3	O2	O1	C10	C9	C8	C7	C6	N5	N3	C5	N4	C4	C3	C2	N2	C1
N1	11	8	6	-7	25	-11	-8	-17	-13	-3	13	-2	-4	-1	1	-5	-3	-2
C1	6	-10	3	-7	17	-11	-16	-22	-10	-6	15	-2	4	-4	1	0	2	
N2	-1	-18	-3	-12	8	-15	-20	-26	-13	-6	6	-7	3	-6	-1	-6		
C2	11	17	10	-2	26	-3	8	0	-5	5	17	-2	11	0	8			
C3	18	25	7	1	36	-3	11	2	-9	1	8	-1	-1	-5				
C4	23	34	18	6	40	0	13	5	-3	0	18	4	-6					
N3	42	25	6	24	57	11	19	10	-1	26	-1	18						
C5	20	-27	-9	10	26	-13	-22	-11	-8	6	-4							
N4	-7	-46	-6	-16	-4	-27	-44	-26	-9	-7								
N5	15	21	15	1	30	-10	1	0	-9									
C6	5	-5	7	-1	10	3	-1	-6										
C7	4	-5	-2	-9	5	-1	-11											
C8	5	-2	18	-6	11	1												
C9	4	11	18	2	1													
C10	-9	3	12	16														
O1	-5	7	20															
O2	-29	-54																
O3	-22																	

Fig. 3 Matrix of differences in mean square displacements [10^{-4} Å^2] along interatomic vectors for the non-hydrogen atoms of adenosine [60]. The shaded region corresponds to the differences between the atoms of the two "rigid" segments of adenosine. The plots of equal-probability ellipsoids for adenosine (50% probability level) correspond to the ADPs estimated using SHADE (*left*) and the values obtained from neutron diffraction experiments, respectively (*right*)

A matrix of such differences for adenosine is given along with plots of equal-probability ellipsoids of measured and estimated ADPs in Fig. 3. The differences are very small between bound atoms in accordance with the "rigid bond test" of Hirshfeld [45], but quite large between the atoms in the adenine and ribose moieties. Accordingly, Klooster and coworkers [60] found that a segmented rigid body model that allows torsional motion about the glycosidic bond N4–C6 was the most satisfactory model. It gives a drop in $R_w(U^{ij})$ from 13.4% to 8.7%. This model was used to test the segmented rigid body approach for estimating hydrogen ADPs [2] as discussed in Sect. 4.4.4.

Once the rigid-body motion of the molecule has been assessed by analysis of the non-hydrogen ADPs it is straight forward to calculate the U^{ij}_{rigid} of the hydrogen atoms based on the rigid body motion. The formulas to perform this calculation can be found in the original literature on the different rigid body analysis approaches.

3.4.2 The Internal Motion

It is well known from Raman and infrared spectroscopic studies that the internal vibrations of hydrogen atoms depend on the chemical environment. First of all, the bond-stretch vibration depends on the type of atom hydrogen is bound to. For example, the stretch-frequency of an O–H bond is much larger than the corresponding frequency of a C–H bond. Second, the functional group influences the bond-bending frequencies, e.g., the out-of-plane vibration of an ethyl group has a larger amplitude than the corresponding vibrations of a methine group. The methods described in the following all assume that these internal vibrations are uncorrelated from the intermolecular or external modes.

The general theoretical framework used to calculate the internal contribution to the ADPs is basically the same whether these estimates are based on information from ab initio calculations, spectroscopic evidence, or neutron diffraction experiments on related compounds. The vibrational modes are expressed in terms of normal coordinates and normal mode frequencies. The modes are uncoupled; each is considered to be an independent harmonic oscillator.

The general relation between vibrational normal mode coordinates and the atomic mean square displacement matrix **B**(k) is [3]:

$$\mathbf{B}(\kappa) = \frac{1}{Nm_\kappa} \sum_{j\mathbf{q}} \frac{E_j(\mathbf{q})}{\omega_j^2(\mathbf{q})} \mathbf{e}(\kappa|j\mathbf{q})\mathbf{e} \times (\kappa|j\mathbf{q})^T, \tag{8}$$

where $e(k|jq)$ represents the kth component of a normalized complex eigenvector $e(jq)$, and corresponds to atom k in normal mode j along the wavevector q. wj is the frequency of mode j and $Ej(q)$ is the energy of the mode, given by

$$E_j(\mathbf{q}) = \hbar\omega_j(\mathbf{q})\left(\frac{1}{2} + \frac{1}{\exp(\hbar\omega_j(\mathbf{q})/k_BT - 1)}\right). \tag{9}$$

In these equations, the energy and frequency of the normal modes depend on the wavevector q. For high-frequency internal molecular vibrations this dependence is negligible, and the equations above can be approximated by

$$\mathbf{B}(\kappa) = \frac{1}{Nm_\kappa} \sum_{j} \frac{E_j}{\omega_j^2} \mathbf{e}(\kappa|j)\mathbf{e}\times((\kappa|j))^T \tag{10}$$

and

$$E_j = \hbar\omega_j\left(\frac{1}{2} + \frac{1}{\exp(\hbar\omega_j)/k_BT - 1}\right). \tag{11}$$

Where **B**(k) is the atomic mean square displacement matrix of atom k. The mean square displacement matrix **B** is expressed in a Cartesian coordinate system and is related to the matrix **U** of ADPs by a similarity transformation, since the ADP matrix is expressed in the generally oblique coordinate system defined by the unit cell axes. The definitions and transformation properties of ADPs can be found in, e.g., the report by Trueblood et al. [4] or the International Tables for Crystallography [28].

The methods differ in the way that the vibrational coordinates and frequencies are obtained and described, and in the number of modes that is used to construct the internal mean square displacement matrix of the hydrogen atoms.

3.4.3 Spectroscopic Evidence

In his early papers, Hirshfeld used information from Raman and infrared spectroscopy to assess the frequencies and corresponding mean square displacements of the bond-stretching and bond-bending modes of the X–H bonds [46, 47]. This approach has been used several times by Destro and coworkers [37, 38, 61–63]. The procedure has been implemented in the code ADPH and has been described and tested in detail by Roversi and Destro [18];

ADPH: In the ADPH approach the normal mode frequencies are based on spectroscopic data and each vibration is described by approximate vibrational coordinates. In the simplest case, three independent modes – one bond-stretching mode and two modes perpendicular to the bond – are used to construct the internal part U^{ij}_{internal} of the internal mean square displacement matrix for the hydrogen atom, however there is no limitation on the number of modes that can be used. This approach has the advantage that the estimates of internal ADPs can be based on experimental evidence on the same compound that is studies by X-ray diffraction. However, for larger molecules with several similar functional groups, it becomes difficult to assign the different spectroscopic frequencies to the right hydrogen atoms, and the approach has to rely on mean group frequencies.

3.4.4 Analysis of Neutron Diffraction Data

It is possible to analyze the vibrational motion of hydrogen atoms in a similar vein as the statistical analysis of X–H bond lengths derived from neutron diffraction studies found in International Tables for Crystallography [39]. When the total atomic mean square displacement matrix U^{ij} has been determined from neutron diffraction experiments, and the rigid molecular motion U^{ij}_{rigid} has been determined from a rigid-body analysis of the non-hydrogen ADPs, it becomes possible to get an estimate of the internal motion of the hydrogen atoms by rearranging (6).

$$U^{ij}_{\text{internal}} = U^{ij} - U^{ij}_{\text{rigid}}. \tag{12}$$

It was noted by Johnson [64] that the mean square displacements derived from U^{ij}_{internal} of hydrogen atoms was in good agreement with spectroscopic information, showing systematic trends corresponding to the functional group that hydrogen was part of. Similar observations were done by Craven and coworkers in the analysis of cholesteryl acetate [65], suberic acid [66], hexamethylene tetramine [67], and piperazinium hexamoate [68]. The internal torsional motion of a range of librating groups, including methyl, carboxyl, and amino groups was also thoroughly investigated by Trueblood and Dunitz [69] based on more than 125 neutron diffraction studies of molecular crystals from the literature.

Madsen and coworkers [11] investigated the internal mean square displacements of hydrogen atoms in xylitol and a range of related carbohydrate compounds found in the literature, and these estimates of internal modes of were collected in a "library" and later improved and enhanced with more statistical material [2]. The present library provides mean values of internal stretch modes as well as in-plane and out-of-plane bending modes for a range of chemical groups involving hydrogen bound to C, N, and O, and forms the basis for assigning ADPs to hydrogen atoms in the *Simple Hydrogen Anisotropic Displacement Estimator* (SHADE) server.

SHADE: The library of internal mean square displacements derived from neutron diffraction studies has been used to construct a web server that allows fast and accurate estimates of the ADPS of hydrogen atoms. The SHADE server [70] allows users to submit a CIF file [71] containing the atomic coordinates and the ADPs of the non-hydrogen atoms. The server performs a TLS analysis using the THMA11 program, and combines the rigid body motion with the internal motion obtained from analysis of neutron diffraction data. It is possible to perform a segmented rigid body analysis using the attached rigid group approach of the THMA11 program [2, 52]. The segmented rigid body approach seems to give marginally better results, as compared to neutron diffraction experiments, as judged from a few test cases [2] on non-rigid molecules. For adenosine we observed that despite the improved description of the motion of the heavy-atom skeleton, only small improvements were observed for the H atom ADPs. For some hydrogen atoms there was a substantial improvement, while for other we observed a worsening agreement. There is definitely room for further testing and improvement of the segmented rigid body analysis in this context.

The SHADE server is available at the web address: http://shade.ki.ku.dk.

3.4.5 Ab Initio Calculations

Estimates of interatomic force-constants obtained from ab initio calculations are today a straightforward way to build the dynamical matrix used in a normal-mode coordinate analysis. Several academic and commercial ab initio codes offer integrated normal-mode analysis. A program XDvib which is part of the charge-density analysis program XD [72] is able to read the output from a normal-mode analysis from Gaussian [73] and compute the ADPs corresponding to internal vibrations. This procedure was used successfully by Flaig et al. [74] to provide ADPs for hydrogen atoms in a study of D,L-aspartic acid. In that study, the external contribution to the ADPs was based on a rigid-link refinement of the non-hydrogen ADPs, which essentially mimics a rigid-body type refinement. However, the ab initio calculation of internal modes was performed on an isolated (gas-phase) molecule. This is not always sufficient to obtain reliable results. Results by Luo et al. [68] and Madsen et al. [11] show that gas-phase calculations can lead to internal mean square displacements that are much larger than the total mean square displacements as derived from neutron diffraction experiments, because the intermolecular potential energy surface (PES) of an isolated molecule is very different from the PES of a molecule in a crystalline environment, especially for non-rigid systems with large amplitude torsional vibrations. The flat PES causes large amplitudes of some of the internal molecular vibrations, e.g., torsional modes. In these cases, it is necessary to take the intermolecular environment into account. For rigid molecules with weak intermolecular interactions it may be sufficient to use gas-phase calculations (e.g., the case of naphthalene [75]). One of the major advantages of the ab initio approach is the possibility of estimating the motion of hydrogen atoms that are in nonstandard environments, i.e., hydrogen atoms in groups that are not well characterized by spectroscopy or neutron diffraction measurements.

TLS+ONIOM: Whitten and Spackman [76] used ONIOM calculations – a procedure where the central molecule is treated using quantum mechanics and a cluster of surrounding molecules is treated using classical molecular mechanics – to mimic the intermolecular environment with excellent results. This "TLS+ONIOM" approach differs slightly from the ADPH and SHADE approaches in that the internal mean square displacements are subtracted from the ADPs of the non-hydrogen atoms before the TLS analysis. Although this is a small correction, it seems to be an improvement as it diminishes the differences between the mean square displacements of bonded atoms in the direction of the bond (this so-called rigid bond test by Hirshfeld [45] is often used to test the reliability of ADPs derived from experiments). An alternative approach is to use a program that is designed for ab initio periodic solid-state calculations. An approach based on the CRYSTAL09 program has been tested by Madsen et al. [77] and gives results in close agreement with the TLS+ONIOM calculations.

3.5 Estimates of H ADPs Based Solely on Force-Field or Ab Initio Calculations

The internal vibrations of hydrogen atoms correspond to the high-frequency part (200–3,500 cm^{-1}) of the Raman and IR spectra. It is well known that these frequencies can be reproduced by ab initio calculations, at least to within a common scale factor (see Scott and Radom [78]) – as witnessed by the success of the TLS+ONIOM approach described above.

However, the low-frequency (0–200 cm^{-1}) external vibrations of hydrogen and non-hydrogen atoms give a significant contribution to the mean square displacement matrix, and this contribution is increasing with temperature, according to (10) and (11). The low-frequency vibrations are more complicated to compute because they correspond to correlated molecular motion and show dispersion, i.e., the frequencies depend on the direction and magnitude of the wavevector. It is possible to estimate these normal modes by using a lattice-dynamical approach, as developed by Born and von Kármán and described in the classical book by Born and Huang [79], and in a language more familiar to crystallographers by Willis and Pryor [3].

The lattice-dynamical evaluation of ADPs based on the Born and von Kármán procedure has been applied for many years based on empirical force-fields. A series of systematic calculations on several rigid molecular crystals, and comparison with experimental results was reported during the 1970s by Fillippini, Gramaccioli, Simonetta, and Suffriti [80–82]. Later results by Criado and coworkers showed that the force-field approach was also sufficient for azahydrocarbons [83]. In these studies it is sometimes found that the calculated ADPs are higher – in some cases by more than 50% – than the experimental results. Gramaccioli and coworkers ascribe this difference to the effect of TDS on experimental ADPs [81, 82], which, because it is seldom corrected for, leads to an increase in the observed diffraction intensities and thereby small ADPs. However, in cases where the force-constants of the Born and von Kármán model are derived from inelastic neutron scattering measurements and infrared spectroscopic measurements, as in the study of urotropine by Willis and Howard [84] and the study of silicon by Flensburg and Stewart [85], there seem to be an excellent agreement with the experimental ADPs, so perhaps some of the discrepancies are due to inadequacies in the applied empirical force fields.

With today's computer power, it has become feasible to perform ab initio calculations in order to assess the force-constants needed for the lattice dynamical treatment. Whereas the ab initio approach seems to work very well for extended solids [86, 87], very little work has been done to test it for molecular crystals. Work in progress [77] on estimation of ADPs for crystalline urea, urotropine, and benzene indicates that the well-known inability of DFT methods to account for dispersive forces may be a major problem for computing the lattice dynamics, and thereby the ADPs, of molecular crystals using DFT methods.

4 Charge-Density Studies with a Special Focus on H Atoms

All charge-density studies of molecular crystals involving hydrogen atoms are to some extent affected by the modeling of the hydrogen atoms. However, some studies focus specifically on the intermolecular interactions, often in terms of hydrogen bonding, or on the properties of the hydrogen atoms, e.g., their electronic properties in terms of the EFGs. In these cases the treatment of hydrogen is of course crucial. Here, we discuss a number of cases where the role of hydrogen atoms have proven to be especially important – and where the authors have considered the modeling of hydrogen carefully – in order to discuss the pros and cons of the different approaches to hydrogen modeling. The list of studies is not meant to present an overview of recent charge-density studies involving hydrogen atoms, but rather a few highlights that will give an impression of limitations and possibilities of different approaches to model and study hydrogen atoms.

4.1 Strong Hydrogen Bonds

Very short and strong hydrogen bonds possess many unique characteristics [1], and they often play an important role in biological systems [88]. Charge-density studies have been used to characterize the electronic environment in the vicinity of the proton involved in strong O–H···O bonds. Is the hydrogen bond symmetric? Does the proton have a unimodal or bimodal probability density distribution – i.e., is it situated in a single or double-well potential? Do the very strong hydrogen bonds resemble the weaker hydrogen bonds or should they rather be characterized as covalent?

MAHS and MADMA: In order to answer these subtle questions, it is important to gather as much information about the hydrogen nuclear parameters as possible. Flensburg and coworkers [13] studied the salt of MAHS – methylammonium hydrogen succinate monohydrate – based on the combined use of neutron and X-ray diffraction data at 110 K. Based on the neutron diffraction data they found that the short O–H···O hydrogen bond was symmetric. A topological analysis of the model of static electron density showed that the hydrogen bond had covalent character; the hydrogen atom formed covalent bonds to both oxygen atoms. These conclusions were confirmed in a subsequent study of MADMA – methylammonioum dihydrogen maleate [14].

Benzoylacetone: Madsen and coworkers [89] carried out a study of the intramolecular hydrogen bonding in benzoylacetone (1-phenyl-1,3-butadione) using 8.4 K X-ray data and 20 K neutron data. In contrast to the symmetrical arrangement of the hydrogen atom found in MAHS and MADMA, the hydrogen atom engaged in the strong hydrogen bond in benzoylacetone is asymmetrically placed in a large flat potential well. A topological analysis of the keto-enol group containing the strong intramolecular hydrogen bond showed that the hydrogen position was stabilized by both electrostatic and covalent bonding contributions at each side of the hydrogen atom.

Isonicotinamide: A recent study of two polymorphs of isonicotinamide and oxalic acid investigated the character of short strong O–H···N intermolecular hydrogen bonds between the acid and the pyridine base [90]. Again, combined X-ray and neutron diffraction techniques were crucial to deconvolute a static electron density from the mean thermal charge density. As for the O–H···O bonds, it was found that the hydrogen bonding had a pronounced covalent character. This study was compared to the charge density obtained from periodic ab initio calculations on the crystalline system.

In these studies where the fundamental questions are related to the exact position and dynamics of the protons, the use of spectroscopic information, or information from ab initio calculations, can provide useful complementary information about the dynamics of the proton, however they cannot provide the detailed information about the mean position of the proton which is crucial for the deconvolution of the dynamics of the static density from the vibrations.

There seems to be no way around using a neutron diffraction approach, in order to get a deconvoluted picture of the motion of the nuclei and the distribution of the electron density. However, if the researcher is prepared to abandon the idea of deconvolution, maximum entropy methods can provide an experimental *vibrationally averaged* electron density, which may answer some of the questions posed previously – does the proton occupy a single or double potential well? How is the strength and covalency of the bonds? Studying the electrostatic potential in the region around the hydrogen bond in question, using a multipole model approach where the hydrogen atom remains unmodeled, may provide some of the same information, as shown in the study by Flensburg and coworkers [13].

In other studies of hydrogen-containing molecules, the focus is a characterization of the intermolecular bonding, not the exact position of a hydrogen atom. In these cases, the estimated hydrogen positions and anisotropic thermal motion might be sufficient, as discussed later (Sect. 5.3).

4.2 EFGs at the H Nuclei

A study that certainly merits attention for its careful treatment of hydrogen atom motion is the charge-density study of benzene carried out by Bürgi et al. [20]. This study demonstrates that neutron diffraction data can be useful even if the data have been collected at other temperatures than the X-ray diffraction experiment. Bürgi and coworkers analyzed neutron diffraction data on benzene collected at 15 and 123 K in terms of a normal-coordinate analysis of ADPs [57, 58]. In this analysis, the temperature-dependent rigid-body motion is separated from the high-frequency internal motion which is temperature independent. In essence, it is a multi-temperature TLS analysis also including parameters to account for the internal motion. The resulting normal coordinate model was then used to derive ADPs for the hydrogen atoms at the desired temperature, which in this case was 110 K. The hydrogen ADPs were then used as fixed parameters in a very elaborate multipole model refined against the X-ray diffraction data. The multipole expansion extended up to the hexadecapole level for the carbon atoms, and up to the quadrupole level for hydrogens as noted by the authors, the latter is essential for a proper description of the deformations of quadrupole symmetry about the H atoms, and crucial in this case; from the model of the static electron density, Bürgi and coworkers were able to derive EFGs at the hydrogen nuclei in quantitative agreement with measurements of nuclear quadrupole coupling constants derived from nuclear quadrupole resonance spectroscopy. This is an important confirmation that the hydrogen nuclear parameters are of an excellent quality. As noted by Brown and Spackman [91], EFGs are sensitive to charge density features involving core electrons which, to be accurately modeled, would require more extensive high-angle diffraction data than are currently available, and a more flexible multipole model. For hydrogen atoms, which lack core electrons, the situation is less prohibitive, but still requires a very elaborate multipole model. As a further confirmation of the quality of the model, the

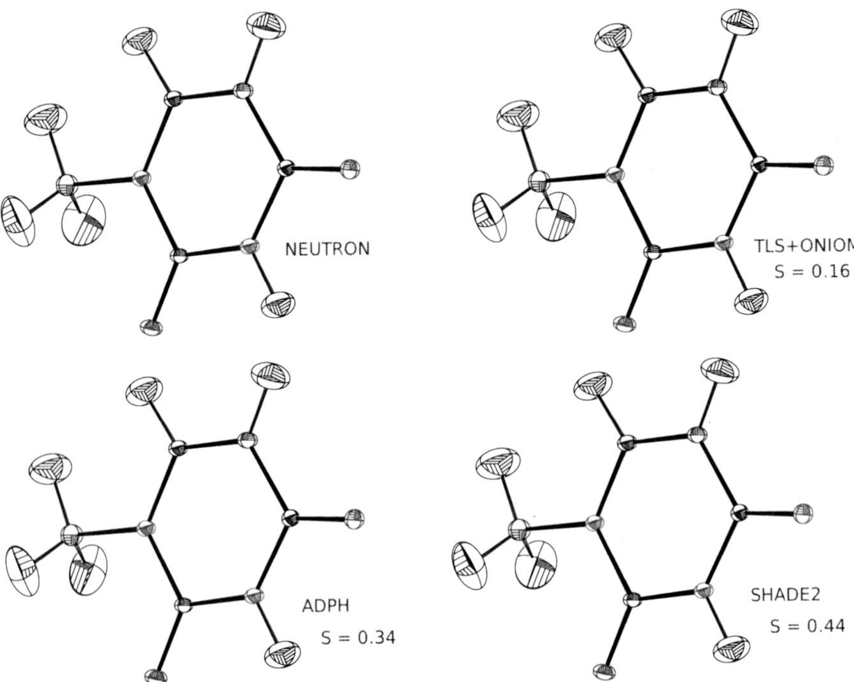

Fig. 4 A comparison of methods to estimate the hydrogen ADPs of 1-Methyl Uracil, based on the work by Munshi and coworkers [2]. To the right of each ellipsoid plot, we give the similarity index [15] between the neutron-derived and estimated ADPs. Equal-probability ellipsoids are shown at the 70% probability level

molecular quadrupole moment derived from the total charge density of the molecule in the crystal also shows to be in excellent agreement with measurements made in the gas phase and in solution. Bürgi and coworkers used neutron diffraction experiments to obtain the hydrogen nuclear parameters. But coordinates and ADPs based on the ADPH approach (Sect. 4.4.3), in conjunction with an elaborate multipole model, seem adequate in order to obtain EFGs at the hydrogen nuclear positions, as judged from the study by Destro et al. on α-glycine [37]. This opens the possibility that also the TLS+ONIOM and SHADE approaches are sufficiently accurate for this type of study (Fig. 4).

4.3 Molecular Interactions

One of the areas where estimated hydrogen ADPs and positions may play an important role is in the study of biologically important molecules, where experimental charge densities are used in the characterization of the electrostatic properties of the molecules and their intermolecular interactions.

Fidarestat: A very recent example is the study of Fidarestat, an inhibitor of the protein human aldose reductase [92]. In this study the hydrogen atoms were restrained to the standard neutron distances as listed in the International Tables for Crystallography [39], and a preliminary multipole refinement was conducted using isotropic displacement parameters and an SDS scattering factor [5] for the hydrogen atoms. Subsequently, the positions and ADPs from this refinement were submitted to the SHADE server [70] in order to estimate ADPs for the hydrogen atoms. The molecule was divided into four rigid groups in the TLS analysis. The hydrogen atom ADPs were then refined using tight restraints to the target values obtained from the SHADE program, and the multipole refinement was then continued. The authors noticed a small, but systematic improvement in the agreement factors on adoption of the anisotropic hydrogen atom description. After inclusion of ADPs for hydrogen the authors did not observe significant unmodeled electron density around the hydrogen atoms, and chose to contract the multipole expansion at the dipole level for the hydrogen atoms. An analysis of the electrostatic potential mapped on the molecular surface (Fig. 5) showed clearly visible

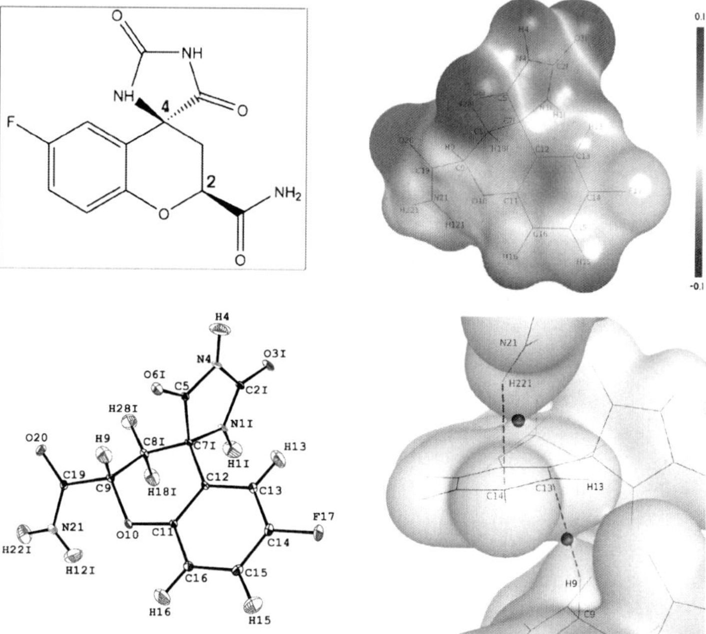

Fig. 5 OrtepIII representation [93] of fidarestat with thermal displacement ellipsoids plotted at 50% probability and the chemical diagram of fidarestat in a frame. Deformation electrostatic potential $\Delta\phi$ (e/Å) generated by the isolated fidarestat molecule mapped on the solvent-excluded surface with a probe radius of 1.4 Å. The potential $\Delta\rho$ was derived from the deformation electron density $\Delta\rho$. The view was made with the program Pymol [94]. View of the hydrogen bonds with π acceptors in a trimer of fidarestat molecules. The aromatic ring containing the C13–C14 atoms is involved on both sides in H···π-system bonds represented as *gray dashed lines*. The electron density cutoff value for the iso-surface is +0.05 e/Å^3. The view was made with the program Pymol [94]

polar binding sites, related to the stronger positive charges of the H–N hydrogen atoms bound to nitrogen atoms compared to H–C atoms. The electrostatic potential pattern was complementary, in a key–lock manner, with the charges of the hydrogen bonded groups in the human aldose reductase active site. A topological analysis of the hydrogen bonding pattern revealed notable $\pi \cdots H\text{–}X$ hydrogen bonds of a strength comparable to C–H···O hydrogen bonds, giving significant contributions to the crystal packing energy.

In this case, the use of estimated hydrogen nuclear parameters seems adequate in order to draw conclusions based on analysis of electrostatic potentials and topological analysis of intermolecular interactions. However, some controversies remain regarding whether it is possible to measure the changes in electron densities due to intermolecular interactions with the present accuracy of experimental charge density studies, as discussed in the following.

Interaction densities: An ongoing debate is whether the quality of present-day experimental charge density studies makes it possible to determine the changes in the charge density due to the interaction with neighboring molecules in the crystal – the interaction density. The charge densities in the intermolecular regions are of a similar magnitude as the typical standard uncertainty of the electron density. A study based on model charge densities obtained from periodic HF calculations on ice VIII, acetylene, formamide, and urea performed by Spackman et al. [95] concluded that the multipole model is capable of qualitative retrieval of the interaction density. The study of Spackman et al. included the effects of thermal motion of the refined electron densities, however no account was taken of the effect of random errors in the simulated structure factors. This effect was included in a subsequent theoretical study of urea by Feil and coworkers [96], which made it impossible to retrieve the interaction density for simulated data.

In a more recent study by Dittrich and Spackman the retrieval of interaction densities was addressed using an experimental charge-density study of the amino acid sarcosine, combined with periodic ab initio calculations using the CRYSTAL98 program [97]. In lack of neutron diffraction data the thermal motion of the hydrogen atoms was based on the TLS+ONIOM approach. Hydrogen positions with bond-lengths in agreement with mean values from neutron diffraction studies were obtained by imposing an electron density model from the invariom database. Dittrich and Spackman conclude that it is possible to retrieve an interaction density from the multipole model of sarcosine, but only if the hydrogen atom electron density was based on the invariom database [98]. The authors note that *it appears that such a highly constrained multipole model is necessary to observe fine details with current data, as the scattering signal of the H atoms is unfortunately rather small when compared to C, N or O atoms.*

Hydrogen bond energies: Closely related to the retrieval of interaction densities is the analysis of intermolecular hydrogen bonding. The standard uncertainty of electron density maps obtained by fitting multipole models against X-ray diffraction data is normally about 0.05 e/Å^3. The electron density of intermolecular bond-critical points is often only a few times higher than this. How reliable is the information obtained from analysis of the electron density in the intermolecular region?

Espinosa and coworkers [99–101] found that the topological properties decay exponentially with the $d_{H \cdots A}$ distance in an analysis of 83 X–H···O interactions from 15 different experimental electron-density studies performed by different investigators using a large variety of models for the hydrogen atoms. However, in a later study of L-histidinium dihydrogen orthophosphate orthophosphoric acid (LHP) [102], it was found that models of hydrogen atoms including quadrupole functions show a quite different behavior – deviating from the exponential dependence – in contrast to models where the multipole expansion is contracted at the dipole level (Fig. 2). The quadrupoles of the H atoms sharpen the electron-density distribution in the plane orthogonal to the H···A hydrogen-bond direction, increasing the perpendicular curvatures and therefore decreasing the Laplacian magnitude at the BCP. Hydrogen-bond interactions that are found as pure closed shell (HCP > 0) in the refinements undertaken without quadrupolar terms present a significant shared-shell character (HCP < 0) when these terms are included. As a consequence, results coming from the models that describe the H atoms up to dipolar terms appear to be in better agreement with theoretical calculations [103] (Fig. 6).

Analysis of intermolecular interactions in epimeric compounds: In a study of the epimeric compounds xylitol and ribitol by Madsen and Larsen [104], it was only

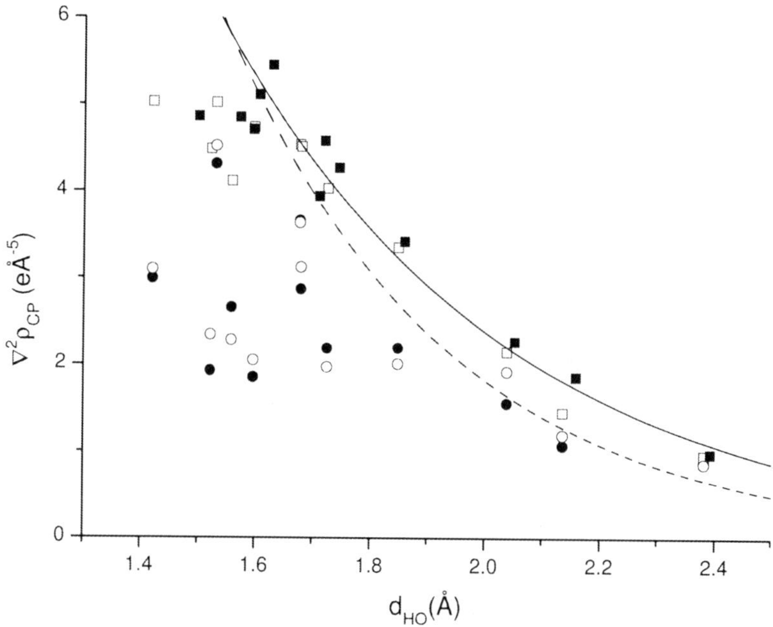

Fig. 6 Topological properties at the BCPs of the hydrogen bonds observed in L-histidinium dihydrogen orthophosphate orthophosphoric acid versus H···O distance. *Open* and *filled circles* correspond to models including quadrupole functions on the hydrogen atoms, *open and filled squares* correspond to models without quadrupole functions on hydrogen. The curves correspond to the empirical dependences by Espinosa et al. [100]. *Solid lines* for X-ray only and *dashed lines* for X-ray+neutron refinements

possible to obtain neutron diffraction data for xylitol because it was impossible to grow crystals of a sufficient size for ribitol. The hydrogen positions for ribitol were based on an IAM refinement and extended to the standard values based on neutron experiments. Hydrogen ADPs were estimated using the SHADE server. A quite elaborate model of the hydrogen electron densities was used, with the multipole expansion extending to the quadrupole level, and with a radial part having a common k parameter for the hydrogen atoms bound to oxygen, and another for the hydrogen atoms bound to carbon. A topological analysis of the intermolecular interactions indicated that the two compounds had identical interaction energies, in agreement with results obtained from calorimetric measurements and periodic DFT calculations. Differences in melting point and mass density could therefore not be explained as a consequence of differences in solid state enthalpies, but were instead related to a difference in solid state entropies, elucidated by TLS analyses of the non-hydrogen ADPs. These results were later confirmed in a more elaborate multi-temperature study [105], where the hydrogen bond energies derived from topological analysis described in Sect. 5.3 are critically discussed as a method of determining the relative stabilities of closely related structures, such as enantiomeric compounds or polymorphs.

Charge density studies including disordered groups: In a recent 85 K X-ray charge density study of paracetamol, Bak and coworkers [106] examine different ways of modeling the disordered methyl group (Fig. 7). In this context, estimated hydrogen ADPs from the SHADE program seemed to offer advantages as compared to the ADPs based on neutron diffraction, partly because of the low quality of

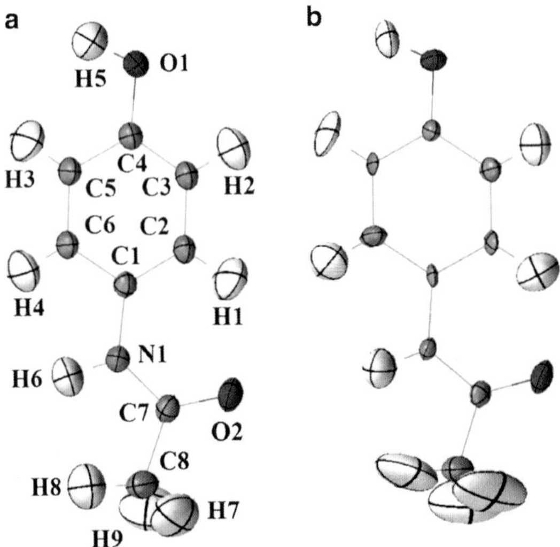

Fig. 7 Structure of paracetamol: (**a**) anisotropic displacement parameters (ADPs) at the 90% probability level for non-H atoms after high-order refinement against X-ray diffraction data (85 K) and for H atoms generated by the SHADE program; (**b**) ADPs at the 90% probability level from neutron diffraction data (80 K; [27])

the neutron diffraction data, and partly because of the disordered methyl groups, which seem to be easier to model using estimated ADPs. The application of ADPs taken from neutron experiments to the methyl H atoms led to extremely high values of electron density at the bond critical points. In the ordered part of the molecule, the use of ADPs generated by SHADE led to better residual maps derived from charge-density models than the neutron ADPs. We note that the other procedures for estimating H ADPs mentioned here (ADPH and TLS+ONIOM) would have been just as efficient as SHADE in this study. This study emphasizes the need to carefully consider the quality of the data. Even when a neutron diffraction dataset is available, it is mandatory to test its validity in every possible way. A high-quality neutron diffraction data set, where the disordered methyl group was carefully modeled using partially occupied sites, would probably have given the most physically reasonable results. Lacking such data, estimated ADPs seem to offer advantages as compared to a mediocre neutron diffraction dataset. A model using isotropic displacement parameters for hydrogen lead to extreme values for molecular electrostatic interaction energies, yet another confirmation that the isotropic model is insufficient for accurate charge density studies.

5 Outlook

Despite the fact that it is now more than three decades since Hirshfeld proposed a method to estimate the anisotropic motion of hydrogen atoms for the use in experimental charge-density analysis, it is only within the last years that the charge-density community at large has become familiar with the method – and with the model defects caused by omitting a careful treatment of the hydrogen atoms. Hopefully, this familiarity has raised general awareness that the description of the static charge density (in terms of the multipole model) and the description of atomic motion (in terms atomic anisotropic motion) are the ying and yang of charge density analysis – both aspects of the model has to be treated carefully, or both will be erroneous. The examples given in the last part of this chapter all reflect an awareness of the coupling between dynamic and static effects with careful treatment of both aspects.

Several tools are now available to estimate the hydrogen ADPs. We have discussed the merits of each of them – as well as advocating the use of neutron diffraction experiments whenever that is possible. However, whereas there seems to be a growing consensus regarding the treatment of nuclear positions and thermal motion for hydrogen, the flexibility of the multipole model of hydrogen is still debated, not only in terms of the truncation of the multipole expansion but also in terms of the flexibility of the radial parameters.

Ab initio calculations have been compared with static charge densities for several years. Looking ahead, the moment seems ripe to compare the dynamics of atoms and molecules obtained from ab initio calculations with the information from careful diffraction experiments.

References

1. Jeffrey GA (1997) An introduction to hydrogen bonding. Oxford University Press, Oxford
2. Munshi P, Madsen AØ, Spackman MA, Larsen S, Destro R (2008) Estimated H-atom anisotropic displacement parameters: a comparison between different methods and with neutron diffraction results. Acta Crystallogr A 64(4):465–475
3. Willis BTM, Pryor AW (1975) Thermal vibrations in crystallography. Cambridge University Press, Cambridge
4. Trueblood KN, Bürgi HB, Burzlaff H, Dunitz J, Gramaccioli CM, Schulz HH, Shmueli U, Abrahams SC (1996) Atomic dispacement parameter nomenclature. Report of a subcommittee on atomic displacement parameter nomenclature. Acta Crystallogr A 52(5):770–781
5. Stewart RF, Davidson ER, Simpson WT (1965) Coherent X-ray scattering for the hydrogen atom in the hydrogen molecule. J Chem Phys 42(9):3175–3187
6. Chandler GS, Spackman MA (1982) Pseudoatom expansions of the first-row diatomic hydride electron densities. Acta Crystallogr A 38:225–239
7. Netzel J, Hofman A, van Smaalen S (2008) Accurate charge density of α-glycine by the maximum entropy method. Cryst Eng Comm 10:335–343
8. Netzel J, van Smaalen S (2009) Topological properties of hydrogen bonds and covalent bonds from charge densities obtained by the maximun entropy method (MEM). Acta Crystallogr B 65:624–638
9. Dietrich H (1976) 'Core electrons' of bonded hydrogen atoms. Acta Crystallogr A 32:347–348
10. Madsen AØ, Sørensen HO, Stewart RF, Flensburg C, Larsen S (2004) Modeling of nuclear parameters for hydrogen atoms in X-ray charge density studies. Acta Crystallogr A 60:550–561
11. Madsen AØ, Mason S, Larsen S (2003) A neutron diffraction study of xylitol: derivation of mean square internal vibrations for hydrogen atoms from a rigid-body description. Acta Crystallogr B 59:653–663
12. Birkedal H, Madsen D, Mathiesen RH, Knudsen K, Weber HP, Pattison P, Schwarzenbach D (2004) The charge density of urea from synchrotron diffraction data. Acta Crystallogr A 60 (5):371–381
13. Flensburg C, Larsen S, Stewart RF (1995) Experimental charge density study of methylammonium hydrogen succinate monohydrate. A salt with a very short O-H-O hydrogen bond. J Phys Chem 99(25):10130–10141
14. Madsen D, Flensburg C, Larsen S (1998) Properties of the experimental crystal charge density of methylammonium hydrogen maleate. A salt with a very short intramolecular O-H-O hydrogen bond. J Phys Chem A 102(12):2177–2188
15. Whitten A, Spackman M (2006) Anisotropic displacement parameters for H atoms using an ONIOM approach. Acta Cryst 62:875–888. doi:101107/S0108768106020787, pp 1–14
16. Bader R (1994) Atoms in molecules. Oxford University Press, Oxford
17. Hoser AA, Dominiak PM, Wozniak K (2009) Towards the best model for H atoms in experimental charge-density refinement. Acta Crystallogr A 65(4):300–311
18. Roversi P, Destro M (2004) Approximate anisotropic displacement parameters for H atoms in molecular crystals. Chem Phys Lett 386:472–478
19. Roversi P, Barzaghi M, Merati F, Destro R (1996) Charge density in crystalline citrinin from X-ray diffraction at 19 K. Can J Chem 74:1145–1161
20. Bürgi HB, Capelli SC, Goeta AE, Howard JAK, Spackman MA, Yufit DS (2002) Electron distribution and molecular motion in crystalline benzene: an accurate experimental study combining CCD X-ray data on C_6H_6 with multitemperature neutron-diffraction results on C_6H_6. Chem A Eur J 8(15):3512–3521
21. Mata I, Espinosa E, Molins E, Veintemillas S, Maniukiewicz W, Lecomte C, Cousson A, Paulus W (2006) Contributions to the application of the transferability principle and the

multipolar modeling of H atoms: electron-density study of L-histidinium dihydrogen orthophosphate orthophosphoric acid. I. Acta Crystallogr A 62(5):365–378
22. Rousseau B, Maes ST, Lenstra ATH (2000) Systematic intensity errors and model imperfection as the consequence of spectral truncation. Acta Crystallogr A 56:300–307
23. Becker PJ, Coppens P (1975) Extinction within the limit of validity of the Darwin transfer equations. I. General formalisms for primary and secondary extinction and their application to spherical crystals. Acta Crystallogr A 31:129–147
24. Becker PJ, Coppens P (1975) Extinction within the limit of validity of the Darwin transfer equations. III. Non-spherical crystals and anisotropy of extinction. Acta Crystallogr A 31:417
25. Morgenroth W, Overgaard J, Clausen HF, Svendsen H, Jørgensen MRV, Larsen FK, Iversen BB (2008) Helium cryostat synchrotron charge densities determined using a large CCD detector – the upgraded beamline D3 at DESY. J Appl Cryst 41(5):846–853
26. Blessing RH (1995) On the differences between X-ray and neutron thermal vibration parameters. Acta Crystallogr B 51:816–823
27. Wilson C (2000) Single crystal neutron diffraction from molecular materials. World Scientific, Singapore
28. Kuhs WF (2003) International tables for crystallography. Kluwer Academic, Dordrecht, pp 228–242
29. Kuhs WF (1992) Generalized atomic displacements in crystallographic structure analysis. Acta Crystallogr A 48:80–98
30. Ruysink AFJ, Vos A (1974) Systematic errors in structure models obtained by X-ray diffraction. Acta Crystallogr A 30(4):503–506
31. Coppens P (1968) Evidence for systematic errors in X-ray temperature parameters resulting from bonding effects. Acta Crystallogr B 24(9):1272–1274
32. Hope H, Ottersen T (1978) Accurate determination of hydrogen positions from X-ray data. I. The structure of s-diformohydrazide at 85 K. Acta Crystallogr B 34:3623–3626
33. Almlöf J, Otterson T (1979) X-ray high-order refinements of hydrogen atoms: a theoretical approach. Acta Crystallogr A 35:137–139
34. Stewart RF, Bentley J, Goodman B (1975) Generalized x-ray scattering factors in diatomic molecules. J Chem Phys 63(9):3786–3793
35. Stewart RF, Spackman M, Flensburg C (1998) VALRAY98 users manual
36. Destro R, Marsh R, Bianchi R (1988) A low-temperature (23-K) study Of L-alanine. J Phys Chem 92(4):966–973
37. Destro R, Roversi P, Barzaghi M, Marsh RE (2000) Experimental charge density of α-Glycine at 23 K. J Phys Chem A 104:1047–1054
38. Destro R, Soave R, Barzaghi M, Lo Presti L (2005) Progress in the understanding of drug-receptor interactions, Part 1: experimental charge-density study of an angiotensin II receptor antagonist (C30H30N6O3S) at T=17 K. Chem A Eur J 11(16):4621–4634
39. Allen FH, Watson DG, Brammer L, Orpen AG, Taylor R (1999) Typical interatomic distances: organic compounds. In: Wilson AJC, Prince E (eds) International tables for crystallography. Kluwer Academic, Dordrect, pp 782–803
40. Overgaard J, Platts JA, Iversen BB (2009) Experimental and theoretical charge-density study of a tetranuclear cobalt carbonyl complex. Acta Crystallogr B 65(6):715–723
41. Steiner T, Saenger W (1994) Lengthening of the covalent O-H bond in O-H...O hydrogen bonds re-examined from low-temperature neutron diffraction data of organic compounds. Acta Crystallogr B 50:348–357
42. Jayatilaka D, Dittrich B (2008) X-ray structure refinement using aspherical atomic density functions obtained from quantum-mechanical calculations. Acta Crystallogr A 64 (Pt 3):383–393
43. Hirshfeld F (1977) Bonded-atom fragments for describing molecular charge-densities. Theor Chim Acta 44(2):129–138
44. Harel M, Hirshfeld F (1975) Difference densities by least-squares refinement. II. Tetracyanocyclobutane. Acta Crystallogr B 31:162–172

45. Hirshfeld F (1976) Can X-ray data distinguish bonding effects from vibrational smearing? Acta Crystallogr A 32:239–244
46. Hirshfeld F, Hope H (1980) An X-ray determination of the charge deformation density in 2-cyanoguanidine. Acta Crystallogr B 36:406–415
47. Eisenstein M, Hirshfeld F (1983) Experimental versus theoretical charge densities: a hydrogen-bonded derivative of bicyclobutane at 85 K. Acta Crystallogr B 39:61–75
48. Cruickshank DWJ (1956) The analysis of the anisotropic thermal motion of molecules in crystals. Acta Crystallogr 9:754
49. Schomaker V, Trueblood KN (1968) On the rigid-body motion of molecules in crystals. Acta Crystallogr B 24:63–76
50. Dunitz JD, Maverick EF, Trueblood KN (1988) Atomic motions in molecular crystals from diffraction measurements. Angew Chem Int Ed Engl 27:880–895
51. Dunitz JD (1978) X-ray analysis and the structure of organic molecules. Cornell University Press, Ithaca
52. Schomaker V, Trueblood KN (1998) Correlation of internal torsional motion with overall molecular motion in crystals. Acta Crystallogr B 54:507–514
53. Spek A (1990) PLATON, an integrated tool for the analysis of the results of a single crystal structure determination. Acta Crystallogr A 46:C34
54. He XM, Craven B (1985) Internal molecular vibrations from crystal diffraction data by quasinormal mode analysis. Acta Crystallogr A 41:244–251
55. He XM, Craven B (1993) Internal vibrations of a molecule consisting of rigid segments. I. Non-interacting internal vibrations. Acta Crystallogr A 49:10–22
56. Capelli S, Hauser J (2004) NKA user manual, 16th edn. Universität Bern
57. Burgi H, Capelli S (2000) Dynamics of molecules in crystals from multi-temperature anisotropic displacement parameters. I. Theory. Acta Crystallogr A 56:403–412
58. Capelli S, Förtsch M, Burgi H (2000) Dynamics of molecules in crystals from multi-temperature anisotropic displacement parameters. II. Application to benzene (C_6D_6) and urea [$OC(NH)_2$]. Acta Crystallogr A 56(5):413–424
59. Rosenfield RE Jr, Trueblood KN, Dunitz J (1978) A test for rigid-body vibrations, based on a generalization of Hirshfeld's 'rigid-bond' postulate. Acta Crystallogr A 34:828–829
60. Klooster W, Ruble J, Craven B, McMullan RK (1991) Structure and thermal vibrations of adenosine from neutron diffraction data at 123 K. Acta Crystallogr B 47(3):376–383
61. May E, Destro R, Gatti C (2001) The unexpected and large enhancement of the dipole moment in the 3,4-bis(dimethylamino)-3-cyclobutene-l,2-dione (DMACB) molecule upon crystallization: A new role of the intermolecular CH…O interactions. J Am Chem Soc 123(49):12248–12254
62. Forni A, Destro R (2003) Electron density investigation of a push-pull ethylene(C14H12NO2 · H2O) by x-ray diffraction at T=21 K. Chem A Eur J 9(22):5528–5537
63. Soave R, Barzaghi M, Destro R (2007) Progress in the understanding of drug-receptor interactions, part 2: experimental and theoretical electrostatic moments and interaction energies of an angiotensin II receptor antagonist (C30H30N6O3S). Chem A Eur J 13(24):6942–6956
64. Johnson CK (1970) Thermal neutron diffraction. Oxford University Press, Oxford
65. Weber HP, Craven B, Sawzip P, McMullan R (1991) Crystal structure and thermal vibrations of Cholesteryl Acetate from neutron diffraction at 123 and 20 K. Acta Crystallogr B 47:116–127
66. Gao Q, Weber HP, Craven B, McMullan R (1994) Structure of suberic acid at 18.4, 75 and 123 K from neutron diffraction data. Acta Crystallogr B 50(6):695–703
67. Kampermann SP, Sabine TM, Craven B, McMullan R (1995) Hexamethylenetetramine: extinction and thermal vibrations from neutron diffraction at six temperatures. Acta CrystallogrA 51:489–497. doi:101107/S0108767394013711, pp 1–9
68. Luo J, Ruble J, Craven B, McMullan R (1996) Effects of H/D substitution on thermal vibrations in piperazinium hexanoate-H11, D11. Acta Crystallogr B 52(2):357–368

69. Trueblood KN, Dunitz JD (1983) Internal molecular motions in crystals. The estimation of force constants, frequencies and barriers from diffraction data. A feasibility study. Acta Crystallogr B 39:120–133
70. Madsen AØ (2006) SHADE web server for estimation of hydrogen anisotropic displacement parameters. J Appl Crystallogr 39:757–758
71. Hall SR, Allen FH, Brown ID (1991) The crystallographic information file (CIF): a new standard archive file for crystallography. Acta Crystallogr A 47:655–685
72. Koritsanszky T, Howard S, Mallinson P, Su Z, Richter T, Hansen NK (1996) XD: a computer program package for multipole refinement and analysis of electron densities from diffraction data, 1st edn
73. Frisch MJ, Trucks GW, Schlegel HB, Scuseria GE, Robb MA, Cheeseman JR, Montgomery Jr JA, Vreven T, Kudin KN, Burant JC, Millam JM, Iyengar SS, Tomasi J, Barone V, Mennucci B, Cossi M, Scalmani G, Rega N, Petersson GA, Nakatsuji H, Hada M, Ehara M, Toyota K, Fukuda R, Hasegawa J, Ishida M, Nakajima T, Honda Y, Kitao O, Nakai H, Klene M, Li X, Knox JE, Hratchian HP, Cross JB, Bakken V, Adamo C, Jaramillo J, Gomperts R, Stratmann RE, Yazyev O, Austin AJ, Cammi R, Pomelli C, Ochterski JW, Ayala PY, Morokuma K, Voth GA, Salvador P, Dannenberg JJ, Zakrzewski VG, Dapprich S, Daniels AD, Strain MC, Farkas O, Malick DK, Rabuck AD, Raghavachari K, Foresman JB, Ortiz JV, Cui Q, Baboul AG, Clifford S, Cioslowski J, Stefanov BB, Liu G, Liashenko A, Piskorz P, Komaromi I, Martin RL, Fox DJ, Keith T, Al-Laham MA, Peng CY, Nanayakkara A, Challacombe M, Gill PMW, Johnson B, Chen W, Wong MW, Gonzalez C, Pople JA (2003) Gaussian 03, Revision C.02. Tech. Rep., Gaussian, Inc., Wallingford, 2004
74. Flaig R, Koritsanszky T, Zobel D, Luger P (1998) Topological analysis of the experimental electron densities of amino acids. 1. D, L-aspartic acid at 20 K. J Am Chem Soc 120:2227–2238
75. Oddershede J, Larsen S (2004) Charge density study of naphthalene based on X-ray diffraction data at four different temperatures and theoretical calculations. J Phys Chem A 108:1057–1063
76. Whitten AE, Spackman MA (2006) Anisotropic displacement parameters for H atoms using an ONIOM approach. Acta Crystallogr B 62:875–888
77. Madsen AØ, Civalleri B, Pascale F, Dovesi R (2012) Anisotropic displacement parameters for molecular crystals from periodic HF and DFT calculations. Acta Crystallogr A (in press)
78. Scott AP, Radom L (1996) Harmonic vibrational frequencies: an evaluation of Hartree-Fock, Moller-Plesset, quadratic configuration interaction, density functional theory, and semiempirical scale factors. J Phys Chem 100:16502–16513
79. Born M, Huang K (1954) Dynamical theory of crystal lattices. Clarendon, Oxford
80. Filippini G, Gramaccioli CM, Simonetta M, Suffritti BG (1974) On some problems connected with thermal motion in molecular crystals and a lattice-dynamical interpretation. Acta Crystallogr A 30:189–196
81. Gramaccioli CM, Filippini G (1983) Lattice-dynamical evaluation of temperature factors in non-rigid molecular crystals: a first application to aromatic hydrocarbons. Acta Crystallogr A 39:784–791
82. Gramaccioli CM, Filippini G, Simonetta M (1982) Lattice-dynamical evaluation of temperature factors for aromatic hydrocarbos, including internal molecular motion: a straightforward systematic procedure. Acta Crystallogr A 38:350–356
83. Criado A (1990) Lattice dynamics and thermal parameters in azahydrocarbons. Acta Crystallogr A 46:489–494
84. Willis BTM, Howard J (1975) Do the ellipsoids of thermal vibration mean anything? – analysis of neutron diffraction measurements on hexamethylenetetramine. Acta Crystallogr A 31:514–520
85. Flensburg C, Stewart RF (1999) Lattice dynamical Debye-Waller factor for silicon. Phys Rev 60(1):284–290

86. Parlinski K, Li ZQ, Kawazoe Y (1997) First-principles determination of the soft mode in Cubic ZrO$_2$. Phys Rev Lett 78(21):4063–4066
87. Lee C, Gonze X (1995) Ab initio calculation of the thermodynamic properties and atomic temperature factors of SiO2 α-quarts and stishovite. Phys Rev B 51(13):8610–8613
88. Langkilde A, Kristensen SM, Lo Leggio L, Molgaard A, Jensen JH, Houk AR, Poulsen JCN, Kauppinen S, Larsen S (2008) Short strong hydrogen bonds in proteins: a case study of rhamnogalacturonan acetylesterase. Acta Crystallogr D Biol Crystallogr 64((Pt 8)):851–863
89. Madsen G, Iversen B, Larsen F, Kapon M, Reisner G, Herbstein F (1998) Topological analysis of the charge density in short intramolecular O-H · hydrogen bonds. Very low temperature X-ray and neutron diffraction study of benzoylacetone. J Am Chem Soc 120(13):10040–10045
90. Schmidtmann M, Farrugia LJ, Middlemiss DS, Gutmann MJ, McIntyre GJ, Wilson CC (2009) Experimental and theoretical charge density study of polymorphic isonicotinamide-oxalic acid molecular complexes with strong O · N hydrogen bonds. J Phys Chem A 113(50):13985–13997
91. Brown AS, Spackman M (1994) The determination of electric-field gradients from X-Ray-diffraction data. Mol Phys 83(3):551–566
92. Fournier B, Bendeif EE, Guillot B, Podjarny A, Lecomte C, Jelsch C (2009) Charge density and electrostatic interactions of fidarestat, an inhibitor of human aldose reductase. J Am Chem Soc 131:10929–10941
93. Johnson CK (1976) ORTEPII. Report ORNL-5138. Tech. Rep., Tennessee
94. Delano WL (2006) The PyMOL molecular graphics system. Tech. Rep., Delano Scientific, San Carlos
95. Spackman M, Byron PG, Alfredsson M, Hermansson K (1999) Influence of intermolecular interactions on multipole-refined electron densities. Acta Crystallogr A 55:30–47
96. de Vries RY, Feil D, Tsirelson VG (2000) Extracting charge density distributions from diffraction data: a model study on urea. Acta Crystallogr B 56:118–123
97. Saunders VR, Dovesi R, Roetti C, Causa M, Harrisonk NM, Orlando R, Zicovich-Wilson CM (1998) Crystal 98 users manual
98. Dittrich B, Spackman M (2007) Can the interaction density be measured? The example of the non-standard amino acid sarcosine. Acta Crystallogr A 63(Pt 5):426–436
99. Espinosa E, Molins E, Lecomte C (1998) Hydrogen bond strengths revealed by topological analyses of experimentally observed electron densities. Chem Phys Lett 285:170–173
100. Espinosa E, Souhassou M, Lachekar H, Lecomte C (1999) Topological analysis of the electron density in hydrogen bonds. Acta Crystallogr B 55:563–572
101. Espinosa E, Lecomte C, Molins E (1999) Experimental electron density overlapping in hydrogen bonds: topology vs. energetics. J Chem Phys 300:745–748
102. Mata I, Espinosa E, Molins E, Veintemillas S, Maniukiewicz W, Lecomte C, Cousson A, Paulus W (2006) Contributions to the application of the transferability principle and the multipolar modeling of H atoms: electron-density study of l-histidinium dihydrogen orthophosphate orthophosphoric acid. I Acta Crystallogr A62:365–378. doi:101107/S0108767306025141, pp 1–14
103. Espinosa E, Alkorta I, Elguero J, Molins E (2002) From weak to strong interactions: a comprehensive analysis of the topological and energetic properties of the electron density distribution involving X–H ··· F–Y systems. J Chem Phys 117(12):5529–5542
104. Madsen AØ, Larsen S (2007) Insights into solid state thermodynamics from diffraction data. Angew Chemie Int Ed Engl 46:8609–8613
105. Madsen AØ, Mattson R, Larsen S (2011) Understanding thermodynamic properties at the molecular level: multiple temperature charge density study of ribitol and xylitol. J Phys Chem A 115(26):7794–7804
106. Bak J, Dominiak P, Wilson C, Wozniak K (2009) Experimental charge-density study of paracetamol – multipole refinement in the presence of a disordered methyl group. Acta Crystallogr A 65:490–500. doi:101107/S0108767309031729, pp 1–11

Struct Bond (2012) 146: 53–74
DOI:10.1007/430_2010_28
© Springer-Verlag Berlin Heidelberg 2010
Published online: 3 November 2010

Charge Density Methods in Hydrogen Bond Studies

Jacob Overgaard and Bo B. Iversen

Abstract The history of hydrogen bonding is briefly outlined and the current descriptive elements used in the understanding and the study of hydrogen bonds are summarized. The specific challenges that emerge when making experimental charge density studies of hydrogen bonded systems are explained and a number of recent and important charge density studies have been selected to illustrate the significant impact that the experimental charge density modeling continues to have on the understanding of the nature of the hydrogen bonding, both in the electrostatic, longer hydrogen bonds and in the very short, strong hydrogen bonds.

Keywords Bond critical point · Energy density · Hydrogen bonds · Low barrier hydrogen bond · Multipole refinement · Source function · Topological analysis

Contents

J. Overgaard (✉) and B.B. Iversen
Department of Chemistry and Interdisciplinary Nanoscience Center (iNANO), Aarhus University, Langelandsgade 140, 8000 Aarhus C, Denmark
e-mail: jacobo@chem.au.dk

1 Introduction

The hydrogen bond is ubiquitous in chemistry and materials science and has been attributed many unique features. In its simplest form, the hydrogen bond (HB) provides a pathway for attractive interaction between two chemical entities consisting on the one side of a hydrogen atom donor (D–H) and on the other an HB acceptor (A). HBs can contribute stabilizing energetic contributions to the system ranging from fractions of kcal/mol to as much as 44.6 kcal/mol for the dissociation energy of FHF^- into HF and F^- [1]; the exact value of the HB energy is strongly dependent on the D···A distance, but many other factors are appropriate to consider as well.

The first mention of the hydrogen atom as a mediator of interaction between two atoms is ascribed to Moore and Winmill [2], and later developed by Latimer and Rodebush in 1920 who proposed a theory suggesting that the hydrogen, or the proton as they denote it, forms a bond between two different atoms [3]. This was confirmed both by structural studies using electron diffraction by Pauling and Brockway [4], and by spectroscopy by Wulf et al. [5, 6]. The significance of the HB in biology and physiology, which will be mentioned later on, was even anticipated by Pauling as he stated that "... I believe that as the methods of structural chemistry are further applied to physiological problems it will be found that the significance of the HB for physiology is greater than that of any other single structural feature" [7]. Pauling expanded his view on the bonding in these systems and concluded that having only one orbital (1s) the hydrogen atom cannot form more than one pure covalent bond and assigns the interaction to be of mainly electrostatic origin, which has recently been repeated [8]. These previous scientific achievements have paved the way to our present-day understanding of the HB.

In recent years, the HB has been established as an ever-present interaction in most molecular crystals, and the attention for some time has been turned toward characterizing and classifying the HB more meticulously based on geometrical considerations [9]. Concurrently with the structural work that forms the basis of this book, spectroscopic approaches have contributed significantly to the description and understanding of the HB. NMR techniques [10] are widely used to estimate the strength of the HB as an increase in the strength of the HB causes a decreased electronic shielding of the hydrogen nucleus and consequently a downfield shift, which for the strongest HBs is around 20 ppm. Other spectroscopic techniques have given similar significant contributions but remain outside the scope of this chapter and will not be discussed further here. The literature is rife with structural correlation studies on hydrogen bonding, which are made possible mainly due to the enormous improvement of X-ray sources for structural studies, in particular the advent of the X-ray sensitive 2D detector. HBs are also prevalent in crystal engineering although it has a strong unfulfilled potential in such applications. The HB is by nature spatially oriented such that it can provide a very efficient, structure-directing building block in the construction of extended structures and molecular network [11, 12]. One of the most active groups in HB research is the one led by Gilli. Among their many contributions [13], they have arranged the stronger spectrum of homonuclear OHO HBs in three different typical categories [14] of which the resonance-assisted hydrogen bond (or RAHB) is particularly interesting, the other types being positively or negatively charge-assisted hydrogen bonds (±CAHBs). For RAHB systems, they established a correlation between the delocalization in the π-system of the keto-enol system and the donor–acceptor distance:

$$\lambda = 3.47(3) - 1.25(10) * \mathrm{d}(\mathrm{O} \cdots \mathrm{O}),$$

where the delocalization parameter, λ, is calculated as

$$\lambda = 0.5 * \left(1 - \frac{Q}{0.32}\right)$$

with $Q = d_1 - d_2 + d_3 - d_4$. (see Scheme 1) [14].

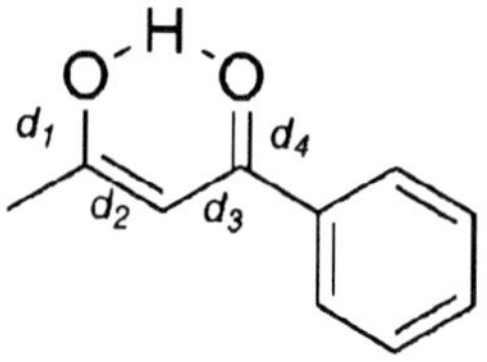

Scheme 1 Bond distances in the antisymmetric vibration parameter, Q, illustrated for benzoylacetone

With the advent of charge density (CD) studies in the late 1970s, the notion of the HB as a purely electrostatic interaction was challenged. Stevens et al. [15] studied sodium hydrogen diacetate that features a very short OHO HB (d(O···O) = 2.45 Å) with the hydrogen positioned on a twofold rotation axis exactly between the two oxygen atoms. The work combined X-ray and neutron data to locate the hydrogen atom and describe its thermal motion and by using deformation density mapping, the authors discovered a significant accumulation of electron density between the hydrogen and both the oxygen atoms suggesting the presence of covalency in both H–O interactions. Thus, this study suggested that the HB is more than a mere electrostatic interaction between two oppositely charged atoms. The next confirmation of covalency in HB was not presented until a decade later by Flensburg et al. [16]. The authors studied the experimental CD in methylammonium hydrogen succinate monohydrate which exhibits a very short O–H···O interaction (d(O···O) = 2.44 Å) with the hydrogen positioned on a symmetry element of the space group. Only the combined use of complementary neutron and X-ray diffraction data made it possible for the authors to conclude that the hydrogen indeed was sitting in a single-well potential, whereas without the neutron data the models with either a single centered hydrogen atom or two half-occupied hydrogen sitting symmetrically away from the midpoint resulted in equally reliable refinements. Similar to the observations by Stevens et al., they found a significant accumulation of electron density in the bond which they quantified by introducing topological analysis of the density.

As foreseen by Pauling, the HB is pervasive in biological systems [7]. The properties of bulk water represent just one of the remarkable examples of the impact of HB as it is responsible for the lower density of solid water compared to liquid water. Another widespread function of HB is found in the active sites of enzymes [17]. It has been postulated, for instance, that the HB in the active site of the serine proteases is something more than a "normal" HB, and instead shows signs of being of a stronger HB type possessing a shortened N···O distance further stabilized in energy terms. The HB within this active site is created by three amino acids termed His57, Asp102, and Ser195 based on the terminology used in the homologous chymotrypsin and is collectively called the catalytic triad, which is an element that remains conserved in serine proteases [18]. The mechanism that regulates the function of the proteases is cleavage of esters or amides by nucleophilic attack on the carbonyl group from the oxygen (Oγ) of Ser195 and creation of a so-called oxyanion hole [19]. The structure of this transition state is the object of speculations; one particular notable theory suggests that it is stabilized by a low barrier HB (LBHB) between the His57 and Asp102 [20]. This proposal has proven rather controversial and provoked heated debates with contributions from several different disciplines however none providing clear-cut evidence in favor of or rejecting the idea [21]. However, very recently a combined high-resolution (in terms of the protein diffraction community which classify 1 Å data set as high-resolution) neutron and X-ray structure of a pancreatic elastase captured in the tetrahedral transition state using an inhibitor appeared [22]. This work showed an N···O separation of 2.60 Å, which may be sufficiently short to exhibit an LBHB; however, the neutron data showed clearly the hydrogen atom located close to the N atom and no sign of an LBHB was evident.

Topological analysis of the electron density, which has been explained in detail in a previous chapter of this book, was early on adopted in the study of hydrogen bonding and most significantly it is marked by the seminal paper by Koch and Popelier [23] in which eight necessary criteria were outlined for an interaction to be called a proper HB. In that work, the authors operated with reference monomer complexes which obviously did not exhibit any intermolecular interactions. Their HB criteria were then formed based on comparison of the monomer properties with hydrogen bonded systems created from dimers of a selection of small molecules. The criteria were as follows:

1. The hydrogen atom is connected to the acceptor atom through a bond path.
2. The value of the electron density at the bond critical point (bcp) is an order of magnitude smaller than what is observed in covalent bonds.
3. The Laplacian of the density gives a small positive value indicative for closed-shell interactions. Both ρ_{bcp} and $\nabla^2{}_{\mathrm{bcp}}{}^2$ correlate with the energy of the HB.
4. The fourth and the last local property to evaluate is the total mutual penetration of the H and A atoms. This penetration depth upon bonding is defined as the difference between the nonbonded radius $\left(r^0_{\mathrm{H,A}}\right)$ of the atom and the bonded atomic radius ($r_{\mathrm{H,A}}$). The latter quantity is the distance from the nucleus to the bcp, while the former is calculated theoretically from a monomer density as the

distance from a perimeter atom to a given CD contour in the direction of the HB. This of course requires that the HB in question is intermolecular otherwise approximations are necessary. The choice of the bcp as the defining point for the bonded atomic radius enables a separation of the total penetration in two atomic terms: $\Delta(r_{\mathrm{H}}) + \Delta(r_{\mathrm{A}}) = (r_{\mathrm{H}}^0 - r_{\mathrm{H}}) + (r_{\mathrm{A}}^0 - r_{\mathrm{A}})$. If this quantity is positive, there is penetration into atomic basins and the HB is formed.

5. The remaining properties deal with integrated atomic properties that are evaluated for the three atoms participating in the HB. The first requirement is the loss of atomic charge from the H atomic basin upon HB formation. The value evaluated is ΔN, the difference in the electronic populations of the monomer and the HB complex, which is negative.
6. The hydrogen atom should be destabilized upon HB formation, such that $\Delta E(\mathrm{H})$ is positive ($E(\mathrm{H})$ is the total atomic energy).
7. The dipolar polarization of the H atom should decrease.
8. The atomic volume decreases upon formation of an HB although this is not set out as a necessary condition.

These conditions for the existence of an HB based entirely on an analysis of the charge distribution remain intact when the interaction in question is not too different from the training set which was used to derive the rules.

A more qualitative approach to the study of HBs followed the work by Abramov [24] on the energetic organization of atomic interactions. The Laplacian is related to the electronic kinetic [$G(r)$ everywhere positive] and potential energy [$V(r)$ everywhere negative] densities through the local virial theorem:

$$\frac{1}{4}\nabla^2\rho = 2G(r) + V(r).$$

The formula implies that the potential energy density dominates in the regions, where the electronic charge is locally concentrated while kinetic energy density dominates in regions with depleted charge density. The disadvantage of the formula is that the evaluation of the energy density is in principle available only from a theoretical wave function. However, when Abramov suggested using Kirzhnits' approximation [25] for $G(r)$ the methodology became available to experimentally determined charge densities:

$$G(r) = \frac{3(3\pi^2)^{2/3}}{10}\rho(r)^{5/3} + \frac{(\nabla\rho(r))^2}{72\rho(r)} + \frac{1}{6}\nabla^2\rho(r).$$

The approximation is only valid in those regions where the density is low and unchanging, i.e., small values of $\rho(r)$ and $\nabla^2\rho(r)$. The impact of this formula was fully developed when Espinosa et al. found a remarkable correlation between the dissociation energy and the potential energy density [26, 27]. The group evaluated a number of previously published HB-containing CD studies and employed the Abramov expression and the virial theorem to calculate energy densities, which revealed an exponential correlation with the hydrogen-acceptor distance (d(O···H)). Furthermore, they calculated dissociation energies (D_e) for HBs from simple dimer theoretical calculations and found a similar exponential correlation between D_e and d(O···H). The comparison of the two fitted exponentials allowed the authors to suggest an extremely simple relationship between the dissociation energy and the potential energy density at the bcp ($E = -D_e = 0.5*V(r_{bcp})$). In a completely different approach [28], van Smaalen et al. have used the maximum entropy method (MEM) to derive charge density distributions for hydrogen bonded systems and calculated the bond energies from this density which is not parameterized but instead exist on a grid. van Smaalen et al. found that the CD from the MEM gives a good agreement with theoretical results.

The next step in the understanding of hydrogen bonding came with the introduction of the source function by Bader and Gatti [29]. They realized that the local source given by

$$LS(r, r') = -\frac{1}{4\pi} \frac{\nabla^2 \rho(r')}{|r - r'|}$$

generates the total electron density by integration:

$$\rho(r) = \int LS(r, r')\mathrm{d}r'.$$

It is obvious to separate the integration of the whole space into integrations over the individual atomic basins which suggest then that the electron density *at any given point* is equal to a sum of contributions from each atomic basin, the contribution being the integration of the local source within the basin. Overgaard et al. [30] showed that there exists a clear correlation between the strength of the HB and the percentage source contribution from both the hydrogen atom and the donor and acceptor atoms. Within a co-crystallized molecular complex, which was intentionally synthesized to be similar to the catalytic triad, three relatively short N–H···O HBs were present which showed a clear trend of the %S(H) to the relevant hydrogen bcp. Incorporating source function data on a few other systems with significantly shorter O–H···O HBs and backed by very detailed theoretical studies by Gatti et al. [31] on hydrogen bonding in small dimeric systems it became clear that the %S(H) started at negative values in weak HBs and increased to give positive contributions for the strongest O–H···O HBs, which then obviously makes the source function a very important indicator for HB strength. The use of the source function has until recently been limited to theoretical wave functions but it can now be calculated also from an experimental multipole density although some limitations still apply. In the same studies, [30, 31] it was found that the relative contributions from the donor and acceptor atoms (%S(D) and %S(A)) increased and decreased, relatively, when the HB distance increased and its strength decreased. The individual contributions from the three atoms (%S(H); %S(A); %S(D)) changed in a concerted manner so that the sum of the three remained relatively unchanged at around 80% in the entire range.

2 Charge Density Studies of Hydrogen Bonds: Requirements and Compromises

The correct description of hydrogen atoms in CD studies has always been a challenge due to the fact that only limited information comes with the X-ray diffraction data used to model it. Naturally this originates in the fact that hydrogen does not possess an unperturbed electronic core, which can be used to establish its nuclear position and possibly also its (anisotropic) vibration amplitude. Instead all X-ray diffraction data comes from the valence electron, which is partly displaced into the covalent bond it shares with its neighboring atom, and a seemingly shortened X–H bond length results. Information about the vibrations of the hydrogen nucleus is only hardly available from X-ray diffraction data and as a consequence it has therefore been common practice in normal structural crystallography to position the atom in a calculated position and constrain its thermal motion to a value derived from the vibrations of its bonded atom. While this approach works reasonably well for a structural model, it is insufficient in the description of the CD.

Madsen et al. showed clearly that using only isotropic atomic displacement parameters for hydrogen atoms had a significant impact on the bcp properties also in bonds not connected to hydrogen within the studied ribose molecule [32]. These observations were found by Hoser et al. to apply to integrated properties as well [33]. In their study, Hoser et al. revisited a number of previously published CD studies and tested different refinement strategies which made them conclude that "to obtain the best topological parameters in the case of a lack of neutron data, a mixed refinement (high-order refinement of heavy atoms, low-angle refinement of H atoms and elongation of the X–H distance to the average neutron bond lengths) supplemented by an estimation of anisotropic thermal motion of H atoms should be applied." Surprisingly, they suggested that this approach is sufficient even in the case of strong hydrogen bonds. It is therefore important to include a more advanced description of the nuclear parameters for the hydrogen atoms. Such information may be provided from single-crystal neutron diffraction data although this is naturally limited by the requirements of crystal volumes exceeding 1 mm^3. However, this restriction may be eliminated when new facilities become operational and crystal sizes down to 0.3^3 mm^3 may be useable at, for instance, the new TOPAZ diffractometer at the Spallation Neutron Source at Oak Ridge National Lab or the IBX station at J-PARC. A different approach to the anisotropic motion of the hydrogen atom is to calculate the vibrational amplitudes either from purely theoretical methods or to use a combined method which involves a rigid body analysis [34] to derive the low-frequency vibrations and adding to this calculated interatomic vibrational amplitudes by categorizing the hydrogen atom according to its bonding environment. This combined method is gaining popularity, particularly since its implementation was made easier by Madsen et al. [35, 36] in the form of a dedicated web-server (SHADE2) that provides the anisotropic parameters for hydrogen atoms based only on a supplied cif file.

A detailed description of hydrogen atomic parameters becomes absolutely essential in studies of strong and very strong HBs in which the position of the hydrogen atom cannot be deduced from the positions of the nonhydrogen atoms. Furthermore, the atomic motion of the hydrogen atom along the line toward the bonding atom is significantly enhanced. In such systems, neutron single crystal data is required to provide the missing information.

The issue of obtainable accuracy from experimental CD studies has recently been critically assessed by Koritsanszky [37].

3 Weak Hydrogen Bonds

The C–H···O HB is a very weak interaction and therefore accurate experimental data are required for its description in electron density studies. Despite its weakness, it is extremely widespread in the solid state and is increasingly used as a structure directing moiety in crystal engineering [38]. Gatti and coworkers [39] studied both the experimental and the theoretical electron density distribution in the crystal structure of the molecule DMACB (Fig. 1), which exhibits a large number of C–H···O contacts exhibiting a large spread of H···O distances while there are no other relevant HBs in the structure. As the work was focused on the properties of the HBs, it is naturally essential to include a proper treatment of these particular atoms and they have therefore been described using anisotropic atomic displacement parameters derived from a rigid-body fit to the non-H atoms as well as implementing supporting information from infrared spectroscopy. The study describes convincingly the angular preference for these weak HBs. First of all, it divides the weak close contacts into bonded and nonbonded interactions based on the existence or not of a bcp between the H and O atoms. With this division made, it is clear that the C–H···O bonds prefer an arrangement with a CHO angle close to 140°. The work

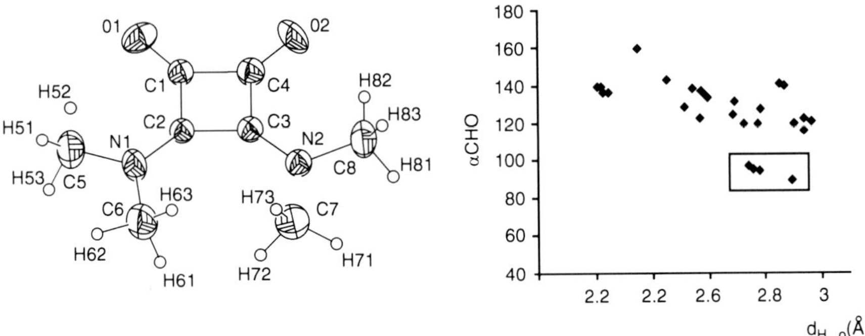

Fig. 1 *Left*: ORTEP drawing of the DMACB molecule. *Right*: The CHO bond angle as a function of the d(H···O) distance. Reprinted with permission from [39]. Copyright 2002 American Chemical Society

also shows the usefulness of the Koch and Popelier criteria [23] and introduces a concept related to intramolecular HBs. For these systems, the penetration of the H and O atoms as well as the atomic properties can change in either direction (positive or negative). This comes about as the hydrogen atoms are already involved in hydrogen bonding in the isolated molecule and the crystallization event can alter the already existing penetration and atomic properties. Therefore, Gatti et al. [39] speak of a *differential* application of the criteria, so that if the penetration is reduced by crystallization then in reality the formation of a crystal reduces the strength of the intramolecular HB. Indeed, the electron population is reduced for all the intramolecular HBs while all the intermolecular HBs clearly obey the normal Koch & Popelier criteria, which require a negative change in penetration. The authors also devote a major part of the paper to a discussion of the reliability of derived dissociation energies from topological properties at the bcps. They use the approximation by Abramov to calculate the kinetic energy density, $G(r_{\mathrm{bcp}})$, at the bcp and from the virial obtain the potential energy density, $V(r_{\mathrm{bcp}})$. Both of these show the expected exponential dependence on the distance from H to O, d(H···O). However, they find that the bond energies derived from equating the energy with half the potential energy density differ significantly from the bond energy derived directly from Espinosa et al. exponential relationship. The former method gives values for the C–H···O hydrogen bond energy up to 18.6 kJ mol^{-1} whereas the latter method provides energies below 9 kJ mol^{-1}. Therefore, Gatti et al. propose that the exponential relationship gives the best results based on the general consensus [40] that the upper bond energy limit for C–H···O HBs is 9.5 kJ mol^{-1} and concluded that the "... behavior of the energy estimate depending on the H···O separation only, seems to rule out the possibility to derive reasonable estimates of the CH···O H-bond energies from the charge density topological analysis or at least to adopt, also in this case, the simple $E_{\mathrm{HB}} = 0.5\mathrm{V}(\mathbf{r_b})$ relationship" [39]. They located the origin of these differences and focused primarily on the fact that the exponential relationship derived by Espinosa could be distorted by the inclusion of a large number of shorter O–H···O and N–H···O HBs while only a relatively few C–H···O HBs were present and not as long as the ones in the DMACB molecule. Accordingly, they derived new parameters for the exponential relationships which were shown to reproduce more dependable bond energies. In fact, when adding up the experimentally derived bond energies from a subset of HBs that exist between two DMACB molecules they find that this total experimental energy is very similar to ab-initio interaction energies for the relevant pair of molecules.

In a related study, Destro and coworkers [41] studied the hydrogen bonding characteristics in the crystal structure of austdiol at 70 K (Fig. 2; the crystal undergoes a number of phase transitions at lower temperatures which is the reason for this unusual temperature) with hydrogen anisotropic displacement parameters evaluated from a rigid-body analysis using the method of Roversi and Destro [42]. Contrary to the above-mentioned study, this structure contains a number of relatively weak competing O–H···O HBs in addition to the C–H···O HBs. What is found – based on the revised exponential equation for the calculation of HB energy from topological properties and bond distance – is that the two types of HBs (C–H···O

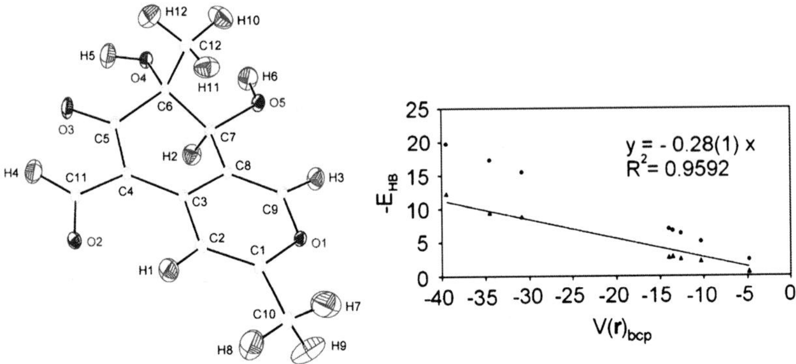

Fig. 2 *Left*: ORTEP drawing of the austdiol molecule. *Right*: The HB energy as a function of the potential energy density at the bcp. Reprinted with permission from [41]. Copyright 2006 American Chemical Society

and O–H···O) behave very similarly in terms of most properties, and that any empirical relationships used in their quantitative classification could potentially cover both types. For instance, the bond angle distribution and the characteristics of the density at the bcp follow the same dependency on the H···O distance.

Destro et al. also discuss the possibility of calculating the electrostatic contribution to the interaction energy in the crystal. They use the widespread energy decomposition scheme by Morokuma and Ziegler [43, 44], which was modified by Spackman to enable its use in experimental CD distributions [45]:

$$E_{\text{int}} = E_{\text{es}} + E_{\text{rep}} + E_{\text{disp}} + E_{\text{pen}},$$

where E_{es} is given by:

$$E_{\text{es}} = E_{\text{pro-pro}} + E_{\text{def-def}} + E_{\text{pro-def}},$$

where pro means the promolecule and def is short for the deformation terms of the molecular charge distribution. They study five different dimers of austdiol (A–E, Fig. 3) and calculate the electrostatic interaction energy (E_{es}) from the experimental charge distribution for all these dimers and compare with theoretical calculations and find that the correspondence between the experimental and theoretical estimates is outstanding. Furthermore, it is clarified, by analysis of the individual components of E_{es}, that the main contributor of attractive electrostatic energy is the promolecule–promolecule energy term. When the total interaction energy (E_{int}) is calculated, the experimental and theoretical values are still in fine correspondence.

The calculation of interaction energies from the (experimental) multipole model is a very recent addition to the available toolkit in the XD program suite [46], and the excellent agreement between theory and experiment shown above suggests that this can successfully be implemented in the study of molecular crystals, and it is

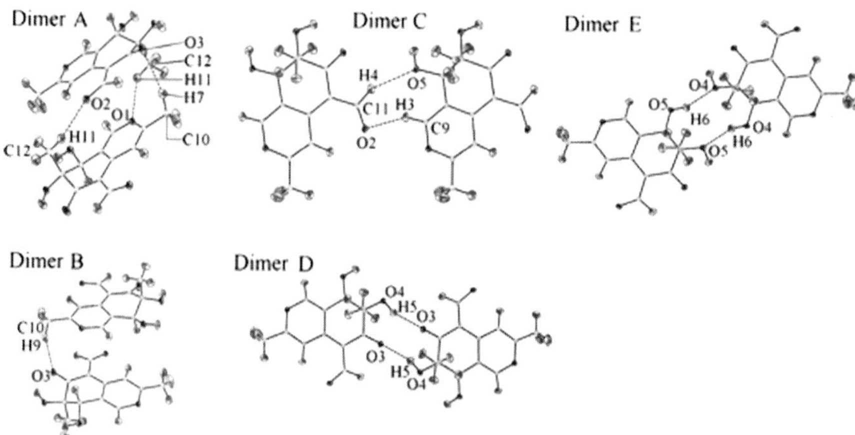

Fig. 3 The five different dimer systems in the austdiol complex used in the calculation of interaction energies from the electron density. Reprinted with permission from [41]. Copyright 2006 American Chemical Society

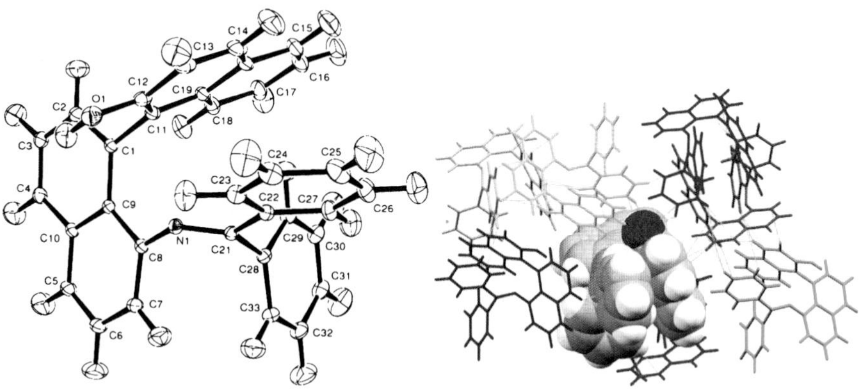

Fig. 4 *Left*: ORTEP drawing of the molecular system. *Right*: The different neighboring molecules in the crystal structure represented by different colors. Reproduced with permission of the International Union of Crystallography from [47]

very likely that there will be more studies of experimental energies in future application of the CD method. One very recent application of the method was performed by Farrugia et al. [47] In this crystal structure (Fig. 4), they found 22 intermolecular interactions of different types – C–H···π(C or N), O–H···π(C), C–H···O, and C–H···H–C. Despite the presence of a hydroxyl group in the crystal and hence the possibility of a stronger O–H···O interaction this is not observed and the abundance of weak interactions can be considered the glue that tie the crystal together. The calculation of the individual bond energies following the work by

Espinosa et al. or the revised formulae provided by Gatti et al. is not performed, but the total interaction energy is calculated using the decomposition scheme presented above in a slightly altered form:

$$E_{\text{int}} = E_{\text{es}} + E_{\text{exc-rep}} + E_{\text{disp}} + E_{\text{ind}},$$

where the induction term is usually neglected as it is inherently accounted for by using the multipole model. The calculation of the seven (Fig. 4, right) dimeric interaction energies from the experimental CD does not provide as good an agreement with the theoretical values as for the austdiol molecule but it is still reasonable.

There are obviously other types of weak HBs; T. N. Guru Row and coworkers have studied the hydrogen bonding of N–H···N, C–H···S, and N–H···S in 1-formyl-3-thiosemicarbazide using a combination of experimental and theoretical CD methods [48]. In the search for the existence of an S–H···N HB, they found this crystal structure but learned that the original structure determination was incorrect and no such interaction was found. Instead, making extensive use of the Koch and Popelier criteria they show that the crystal structure contains seven intermolecular interactions that all fulfill these rules. Without calculating the bond energies in any of the given forms, they conclude that the C–H···S is the weakest of the HB types followed by C–H···O, N–H···N, and N–H···S based entirely on the energy density. Keeping in mind the conclusions from the work of Gatti et al. [39] it is noteworthy that they find the bond energy to be dependent not only on the potential energy density but also on the HB distance. That correlation was based entirely on C–H···O interactions and no assumption was made that it was directly transferable to interactions with different donor and/or acceptor atom types. Nevertheless, considering that the order of the H···X bond length in the seven HBs outlined in 1-formyl-3-thiosemicarbazide do not strictly follow the order of energy densities, a calculation of the bond energy from some empirical relationship would have supported the statement.

In two recent studies, the HB acceptor properties of two completely different chemical moieties have been clarified. The oxonium ion is known not to possess any significant hydrogen acceptor ability and Lyssenko et al. [49] examined the CD in potassium oxonium bis(hydrogensulfate) from 120 K single-crystal diffraction data in an attempt to discover why. The oxonium ion interacts with the sulfate groups creating medium strength O–H···O HBs but there are no HBs with O_{ox} as acceptor. An analysis of the atomic charge of this atom, and the entire oxonium group, does not suggest that the electrostatics should prevent the oxygen from playing the part of an acceptor. On the other hand, their visualization of the 3D Laplacian distribution suggests that the single lone pair on O_{ox} is very compact which can also be inferred from the HOH bond angles that are found to scatter around 111–113° from a literature search. In conclusion, they argue that the compactness of the lone pair, which they also show using the electron localization function (ELF) distribution, makes it nearly inaccessible for hydrogen atoms.

Beckmann et al. studied the CD in a siloxanol (5-dimethylhydroxysilyl-1,3-dihydro-1,1,3,3-tetramethyl-2,1,3-benzoxadisilole) from 240 K synchrotron X-ray data [50]. The study originated in the lack of basicity of the siloxane linkage and they discovered by theoretical calculations that this basicity is increased when the Si–O–Si angle is reduced from the normal range of 135–180° to lower values comparable to C–O–C angles in ether compounds. The CD of the siloxanol crystal structure reveals a relatively weak HB (d(O···O) of 2.87 Å), although they classify it as strong based on the Koch and Popelier criteria. The ELF is derived from wavefunction fitting to the experimental structure factors using the TONTO program by Jayatilaka [51]. This suggests that the nominal two lone pairs on the O atom in the siloxane linker have merged into one broad maximum, although this is not confirmed by a look at the Laplacian distribution. The conclusion is that the decreased bond angle in the siloxane linker is responsible for the increased basicity and they suggest to use this knowledge in the design of new silicon based crown ethers, where it would be advantageous to have more HB acceptors available for structural control.

4 Strong Hydrogen Bonds

One of the most thoroughly studied examples of a RAHB is benzoylacetone [52, 53] (BA), which also represents the first experimental verification of a low-barrier hydrogen bond (LBHB). BA represents a molecular system with close matching of the p*K* values of the donor (pK_a) and the acceptor oxygen (pK_b). The idea of p*K* matching has been investigated particularly using theoretical methods and described as a prerequisite for the existence of short, strong HBs [54]. In the original CD study of BA, large formal charges were found on the hydrogen and the oxygen atoms and this was used to introduce a modified RAHB model, which includes a charge feedback mechanism. The CD study of the hydrogenated BA was completed with the recently published study of deuterated benzoylacetone (DBA) showing a clearly bimodal probability density function (pdf) (see Fig. 5) [55]. It was explicitly demonstrated using theoretical calculations of test potential curves that a weakly asymmetric potential curve in the O···O regime with a barrier lower than the hydrogen zero-point vibrational energy (ZPVE) but higher than the corresponding ZPVE for a deuteron. However, the deuteron was shown to exhibit tunneling between the two positions and the bimodal distribution was suggested in fine accordance with the experimental findings. Very importantly, the application of the Hirshfeld rigid-bond test [56] to the neutron structure showed no signs of disorder, which was, nevertheless, used as explanation by Gilli et al. [13] for the failure of BA to meet the λ-test (see later).

The DBA study also included a rationalization for the presence of a surprisingly asymmetric very strong HB in nitromalonamide (NMA) [57]. The potential curve is essentially similar to the one observed in BA but the shorter O···O separation changes the pdf into a unimodal although distorted curve. The asymmetry in the

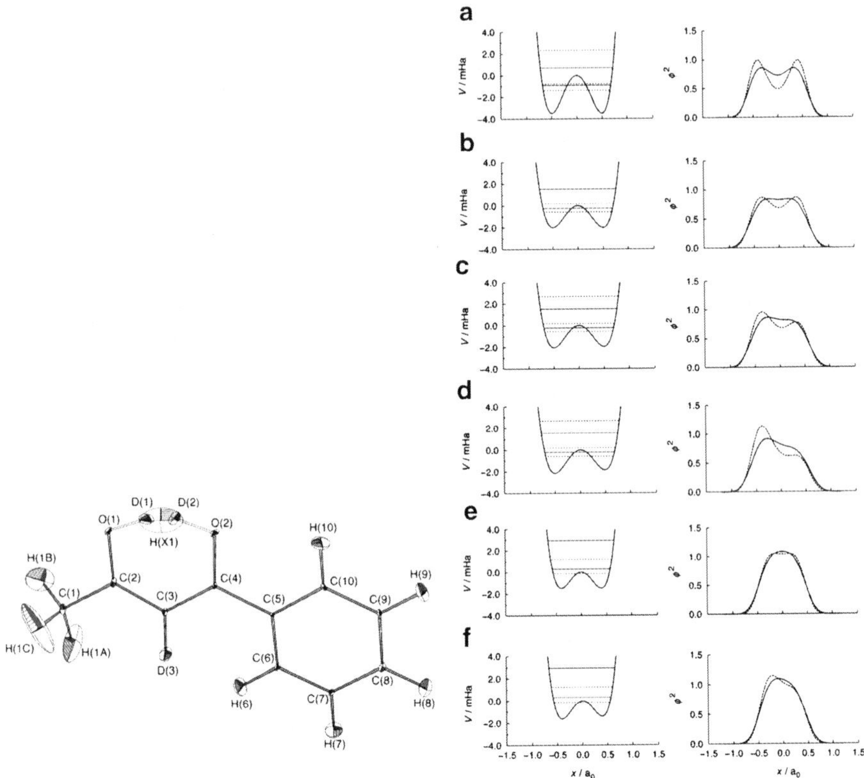

Fig. 5 *Left*: ORTEP drawing of DBA from 20 K neutron diffraction data ($\sin(\theta)/\lambda_{max} = 1.1$ Å^{-1}) showing the two positions of the deuterons, and including also the hydrogen thermal ellipsoid from the neutron study of HBA. *Right*: Potential curves calculated with different degrees of asymmetry (*left column*) and their corresponding pdfs (*right column*). Copyright Wiley-VCH Verlag GmbH & Co. KGaA. Reproduced with permission from [55]

potential curve is explained by asymmetrical intermolecular interactions and the energy differences between a symmetrical and an asymmetrical position of hydrogen is only 0.3% of the total HB energy, which is within the range of a weak additional HB between either the donor or the acceptor atom and an HB donor from a neighboring molecule.

It seems obvious that the hydrogen nuclei in these strong HBs exhibit a vibrational amplitude that is much higher than for the surrounding atoms. Therefore, the probability of finding the hydrogen is nonvanishing even close to the acceptor atom, which suggests that the hydrogen atom sits in a very shallow potential energy minimum. Thus, anharmonic parameters may be important in an accurate description and any meaningful description of the hydrogen location should include also its displacement from the equilibrium position.

More recently, Piccoli et al. have performed a multitemperature neutron diffraction study on tetraacetylethane (TAE, Fig. 6) and determined the CD distribution at

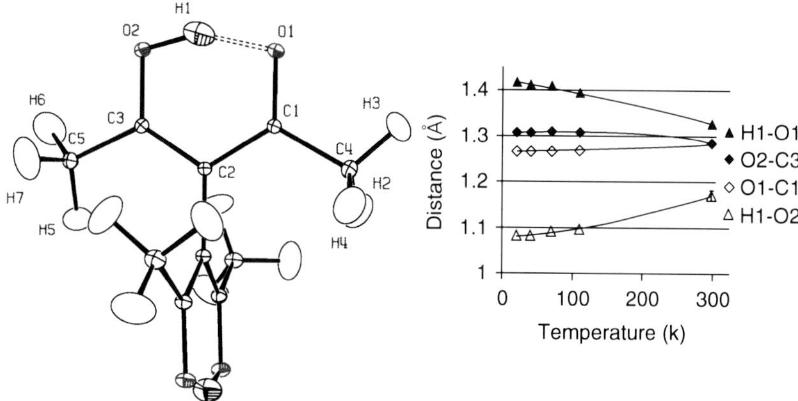

Fig. 6 *Left*: The molecular structure of TAE at 20 K from neutron diffraction data (sin(θ)/λ_{max} = 1.0 Å^{-1}). The HB is shown as a *thick dotted line*. *Right*: The temperature dependence of selected bond lengths. Reprinted with permission from [58]. Copyright 2008 American Chemical Society

a temperature of 20 K [58]. Unfortunately, they have not given any account of whether anharmonic parameters were tested for the hydrogen at the center of the strong HB but the presented model is achieved within the harmonic approximation. The molecular structure of TAE can be viewed as a pK_a matched system as the aliphatic substituents are identical. Nevertheless, there is no imposed crystallographic symmetry on the position of the hydrogen, i.e., it is not restricted to sit at the O···O midpoint. The Hirshfeld rigid-bond test applied to the structures above 110 K indicates the presence of static or dynamic disorder. However, below this temperature the OCCCO moiety behaves as a rigid body which rules out disorder. The hydrogen position changes significantly as the temperature is lowered, from being almost centered at room temperature to being closer to O2 at lower temperatures. The temperature dependence is shown in Fig. 6. The movement of the hydrogen atom position toward the donor atom with decreasing temperature is evidence of a weakly asymmetric, anharmonic potential controlling the hydrogen motion. Given that the O···O distance is much shorter in TAE (2.434(1) at 20 K) than in HBA (2.502(1) Å at 9 K) but the hydrogen position is significantly off-center in this very symmetrical molecule suggests that interatomic interactions act to remove the symmetry. Indeed, there are three short intermolecular C–H···O interactions with O1 or O2 and these can be rationalized to distort the potential and move the hydrogen away from the central position. The intermolecular interactions and the associated asymmetry in the crystal structure of TAE is illustrated using a Hirshfeld surface plot (not shown here). In this respect, TAE strongly resembles NMA which also has a short O···O separation and is a symmetrical molecule but nevertheless exhibits a very asymmetrical, but shallow potential well.

The nature of the hydrogen bonding in TAE was examined using a topological approach of the total electron density and it was found that the HB, i.e., the longer

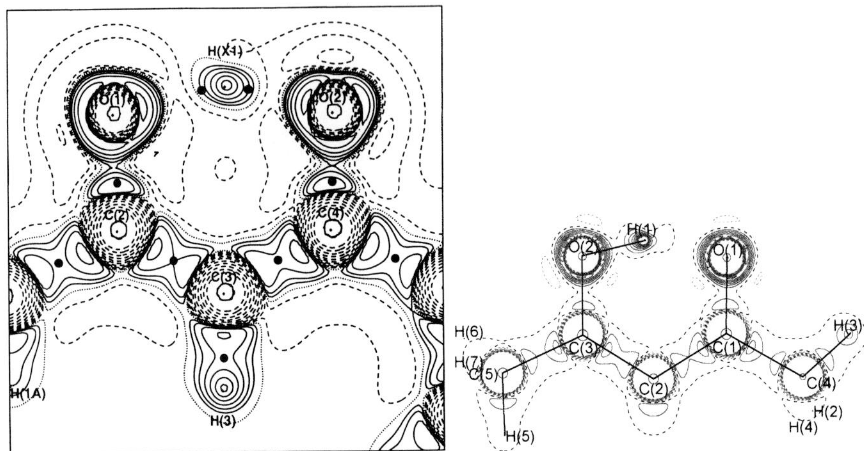

Fig. 7 *Left*: The Laplacian distribution in BA, the *black dots* indicate the positions of the bcps. The contours are drawn at logarithmic intervals of 1.0×2^N eÅ^{-5}. The *dotted line* is the zero contour. *Solid lines* are positive contours, *broken lines* negative contours. The first two positive and negative contours are omitted for clarity. *Right*: The Laplacian distribution in TAE with focus on the HB. Reprinted with permission from [58]. Copyright 2008 American Chemical Society

of the HO interactions, also has a negative value of the Laplacian at the bcp suggesting that it has some covalent character. Both for TAE and HBA the density at the bcp is rather high, approaching values of 1 eÅ^{-3}, which is not much less than what is found in normal covalent bonds. It is also noteworthy that the hydrogen valence shell charge concentration (VSCC) is separated from the oxygen atoms but the bcp is just within the hydrogen VSCC to make the Laplacian negative at the bcp (Fig. 7).

The delocalization of π-density in the OCCCO fragment of a RAHB system can be summarized with the parameter Q and then reduced to a λ-parameter using the empirical relation given by Gilli ($2\lambda_{\mathrm{exp}} = (1-0.32/\mathrm{Q})$). As mentioned above, Gilli correlated the value of λ_{calc} with the O···O separation and found a linear relationship. However, neither BA nor TAE follow the trends from this correlation; BA is much more delocalized that expected from the O···O distance while TAE is much less delocalized than expected from d(O···O). Therefore, these prominent examples expose the deficiencies that accompany structural correlation in strong HB systems based largely on results derived from conventional X-ray data without neutron data to ascertain the exact position of the hydrogen atom. The position of the hydrogen nucleus in strong HBs cannot be inferred from the delocalization within the RAHB fragment; instead knowledge of the closest intermolecular neighbors is imperative – the energy difference between a hydrogen in a central and off-central position is diminutive compared to the energy of the HB and is less than the interaction of a weak, electrostatic CHO HB. Thus, it is, e.g., plausible that different polymorphs of the same strong RAHB system would show different hydrogen atom position and temperature dependence.

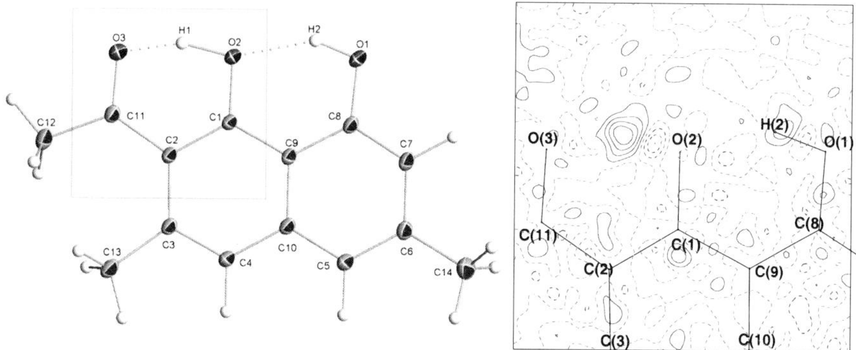

Fig. 8 *Left*: ORTEP drawing of ADD. *Right*: A residual density map calculated without the hydrogen in the model. The contours are shown at 0.1 eÅ^{-3} intervals, *solid lines* show positive and *dashed lines* show negative contours. Reprinted with permission from [59]. Copyright 2007 American Chemical Society

Another recent example showed one of the shortest O···O distances ever seen. In the β-diketone 2-acetyl-1,8-dihydroxy-3,6-dimethylnaphtalene (ADD, Fig. 8), two O–H···O fragments exist with one O···O distance of 2.4031(7) Å and the other 2.6073(7) Å. The crystal structure was determined from very low temperature (15 K) synchrotron X-ray diffraction data but unfortunately the crystal size could not be sufficiently increased to enable a successful neutron diffraction experiment and thus the experimental CD could not be determined, although the residual density map calculated without the hydrogen inserted gives some indication of the position of the hydrogen atom (Fig. 8) [59]. Nevertheless, the study casts significant doubt on the validity of the RAHB theory to describe hydrogen bonded systems as this, where there is much less delocalization in the OCCCO fragment surrounding the hydrogen atom. The value of λ_{exp} is only 0.319 while the calculated value from the empirical correlations by Gilli predicts a completely delocalized system ($\lambda_{calc} = 0.466$). This again exposes the deficiencies by the correlation method and it is not a sensible practice to use the value of λ_{calc} to estimate the position of the hydrogen atom or even less its atomic vibrations. The study relied instead on gas-phase theoretical calculations of the molecule which seems reasonable as the intermolecular interactions in the crystal structure are weak. From these results, it is argued that the leftmost HB, which is the shorter one, is a short strong, low-barrier or single well HB with significant covalent contribution in the H···O interaction. This is based on several important findings: (a) the nonhydrogen skeleton behaves as a rigid body; (b) the hydrogen atom contributes significantly as a source to the density at the H···O bcp; (c) transition state theoretical calculations suggest that the energy of the lowest vibrational state is well above the barrier for hydrogen transfer, making this essentially a single-well HB with a spread out pdf; (d) difference Fourier maps show a nearly centered excess density distribution. The authors suggest that two counteracting forces are in play – on the one side the delocalization of the OCCCO fragment will as part of the RAHB mechanism lower the energy locally but increase the energy of the naphthalene system and thus the effect is that the whole system is in equilibrium without having a completely delocalized RAHB system. Therefore, it seems that the delocalization is not a necessary requirement for the existence of a short, strong HB in a neutral chemical environment such as β-diketones. Another mechanism that plays a role in the creation of the strong HB interaction is steric strain which by the presence of a methyl group at C3 forces O3 closer to H1 than it would be without the methyl group presence. However, theoretical calculations show that even without the presence of this constraint on the geometry of the system, the O···O distance remains short enough to be characterized as a very short HB.

Very recently, the charge density distribution in the crystalline state of two polymorphs of the 2:1 cocrystal of isonicotinamide-oxalic acid (IOA) was published [60]. The work included a combination of neutron (for II the resolution was relatively low due to a small crystal while the data for compound I were of high resolution) and X-ray single crystal diffraction data collected at temperatures of 100 K. The crystal structures (the two molecules from both polymorphs are shown in Fig. 9) feature in both systems a very short heteroatom HB of the type O–H···N with O···N distances of 2.5587(16) Å and 2.539(3) Å, respectively, for forms I and II. The crystallographic statistics suggests that the data quality is high. The positional and thermal parameters for the hydrogen atoms have been adapted by scaling from the neutron parameters, and although there is no indication of whether or to what extent the two independent (neutron and X-ray) diffraction experiments resulted in matching thermal parameters for the nonhydrogen atoms, this is the most favorable procedure.

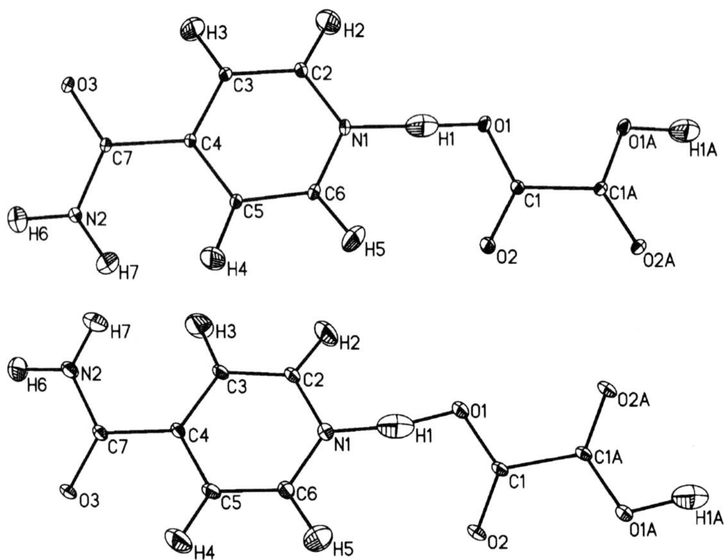

Fig. 9 ORTEP drawings of the two polymorphs of the cocrystal of IO, above is form I and below is form II. Reprinted with permission from [60]. Copyright 2009 American Chemical Society

The study highlights a number of conspicuous points; the atomic probability distributions for the hydrogen atoms in the short, strong HBs are only slightly more extended along the N···O lines. This is interpreted as to not "... indicate any pronounced H mobility within the SSHBs." This is in contrast to the very pronounced mobility of the hydrogen atom observed in the short HB in BA, as also noted by the authors: "To our knowledge, the only case showing a significant H mobility at low temperature was the strong intramolecular O···H···O HB in benzoylacetone." As this review has hinted, only a very limited number of CD studies of short, strong HBs exist and the explanation for the observed discrepancy in the atomic pdf between BA and the remaining structures must necessarily be attributed to the very nature of the HB; the low barrier nature of the HB in BA enhances the proton dynamics significantly and makes the potential considerably more anharmonic. The authors present some calculated potential energy curves which are all significantly anharmonic but nevertheless to some extent provide atomic pdf and

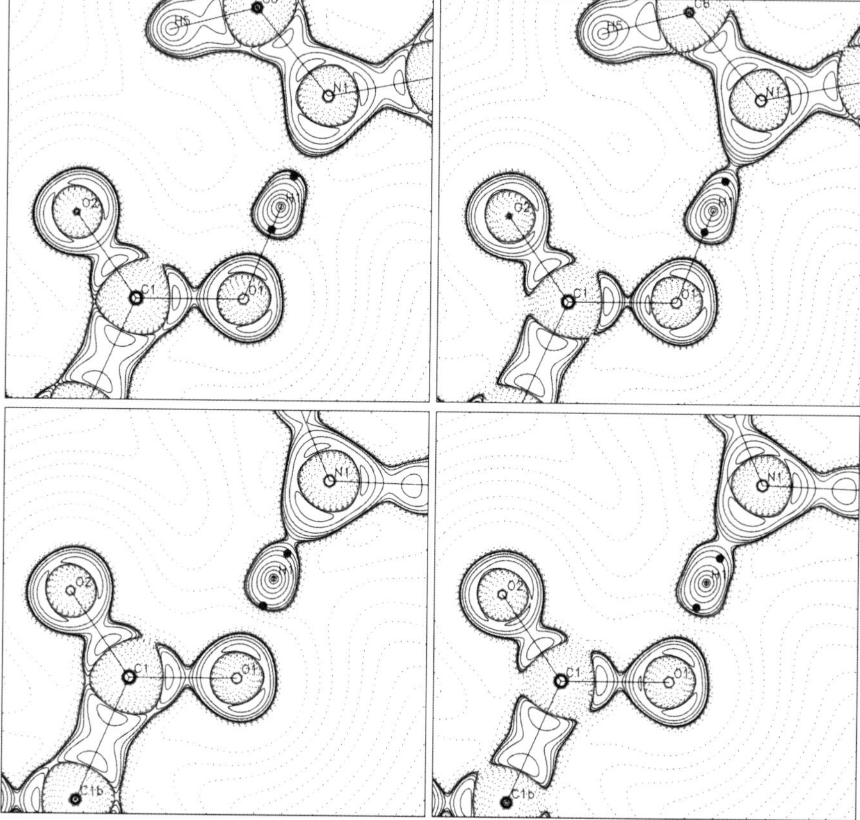

Fig. 10 Theoretical (*left*) and experimental (*right*) Laplacian maps of the short HBs in IOA. Positive contours – *solid black*; zero levels – *solid gray*; negative – *dotted*. Reprinted with permission from [60]. Copyright 2009 American Chemical Society

bond distances comparable to experiment. Based on these considerations, it is therefore not unlikely that the inclusion of anharmonic parameters in the description of the hydrogen in the HB could lead to significant parameters, given also the relatively high temperature of 100 K used in this work in comparison with other studies of short, strong hydrogen bonds.

Another very interesting feature in IOA is the Laplacian distribution (Fig. 10). In particular, form II has a nearly centered hydrogen with O–H and N–H distances of 1.235(5) Å and 1.313(6) Å, respectively. The Laplacian at the bcps in the short HBs are negative as expected for all four interactions in the two forms. However, the charge concentration around the hydrogen merges with the VSCC of the nitrogen atoms both from theory and experiment, which suggests that the bond is perhaps more appropriately described as a N–H···O, more so for form II than form I.

Additional evidence for the strength and nature of this HB comes from the source function, which is illustrated in Fig. 11. First of all, the combined source contributions from the H, A, D triplet is around 80% as expected for this type of strong HB but the nitrogen contributes much more than the oxygen, suggesting that the hydrogen has shifted from the oxygen to the nitrogen. Theoretical calculations indicate bond energies around 100 kJ mol^{-1}, which are not derivable from the bcp properties.

As already mentioned above, there is a different type of potentially strong HB, called the charge assisted HB, or CAHB [61]. There has been some controversy concerning the stabilizing energy of this interaction as Braga et al. questioned the stabilization of the interaction and instead advocated that they should more appropriately be called pseudo HBs [62]. The reasoning behind this proposal was the fact that the CAHB HB brings two negatively charge atoms into proximity and thereby increases the repulsive electrostatic energy. The suggestion was rejected by Steiner

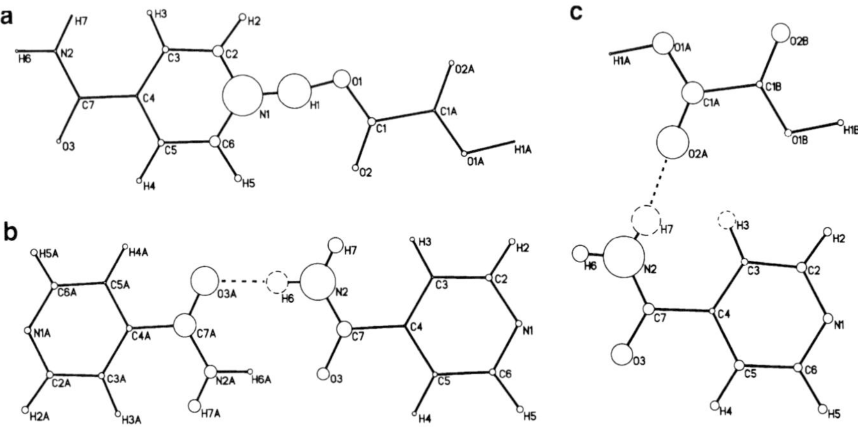

Fig. 11 The source function contributions to three different HBs in form II shown as *circles* with radii scaled according to the value of the contribution. Negative values are shown with *dotted circles*. (**a**) The short O···H···N, (**b**) and (**c**) show the weaker N–H···O HBs. Reprinted with permission from [60]. Copyright 2009 American Chemical Society

in his structure correlation work on HBs [63]. Instead, he expected these HBs to exhibit some degree of covalency. At the same time, Macchi et al. showed using combined neutron and X-ray diffraction data on KHC_2O_4 collected at 10–15 K that the CAHB HB in this system is indeed a rather strong bond [64]. The Laplacian at the H···O bcp is found to be slightly positive but there is a significant polarization of the Laplacian of both the hydrogen atom and the oxygen lone pair involved in the HB. Further evidence for this being a legitimate chemical bond comes from the energy density properties at the bcp; the total energy density is significantly negative while the kinetic energy density is diminished compared to expected values from the H···O distance. Macchi et al. therefore concluded that "In view of the experimental observations reported here and using a more adequate definition of a chemical bond, we find no basis for claiming the break-down of the strength/ length analogy..." which shows the potential of the charge density method in HB studies.

References

1. Wenthold PG, Squires RR (1995) J Phys Chem 99:2002–2005
2. Moore TS, Winmill TF (1912) J Chem Soc 101:1635–1676
3. Latimer WM, Rodebush WH (1920) J Am Chem Soc 42:1419–1433
4. Pauling L, Brockway LO (1934) PNAS 20:336–340
5. Wulf OR, Liddel U (1935) J Am Chem Soc 57:1464–1473
6. Hilbert GE, Wulf OR, Hendricks SB et al (1935) Nature 135:147–148
7. Pauling L (1960) The Nature of the Chemical Bond. Cornell University Press, Ithaca
8. Jeffrey GA, Saenger W (1991) Hydrogen Bonding in Biological Structures. Springer, Berlin
9. Jeffrey GA (1991) An Introduction to Hydrogen Bonding. Oxford University Press, New York
10. Grech E, Klimkiewicz J, Nowicka-Scheibe J (2002) J Mol Struct 615:121–140
11. Ward MD (2009) Struct Bond 132:1–23
12. Desiraju GR (1995) Angew Chem Int Ed 34:2311–2327
13. Gilli P, Bertolasi V, Ferretti V et al (2000) J Am Chem Soc 122:10405–10417
14. Gilli P, Bertolasi V, Ferretti V et al (1994) J Am Chem Soc 116:909–915
15. Stevens ED, Lehmann MS, Coppens P (1977) J Am Chem Soc 99:2829–2831
16. Flensburg C, Larsen S, Stewart RF (1995) J Phys Chem 99:10130–10141
17. Cleland WW, Kreevoy MM (1994) Science 264:1887–1890
18. Kraut JA (1977) Rev Biochem 46:331–358
19. Menard R, Storer AC (1992) Biol Chem Hoppe-Seyler 373:393–400
20. Frey PA, Whitt SA, Tobin JB (1994) Science 264:1927–1930
21. Warshel A (1998) J Biol Chem 273:27035–27038
22. Tamada T, Kinoshita T, Kurihara K et al (2009) J Am Chem Soc 131:11033–11040
23. Koch U, Popelier PLA (1995) J Phys Chem 99:9747–9754
24. Abramov Y (1997) Acta Crystallogr Sect A 53:264–272
25. Kirzhnits DA (1957) Zh Eksp Teor Fiz 5:64–71
26. Espinosa E, Molins E, Lecomte C (1998) Chem Phys Lett 285:170–173
27. Espinosa E, Souhassou M, Lachekar H et al (1999) Acta Crystallogr Sect B 55:563–572
28. Hofmann A, Netzel J, van Smaalen S (2007) Acta Crystallogr Sect B 63:285–295
29. Gatti C, Bader RFW (1998) Chem Phys Lett 287:233–238
30. Overgaard J, Schiøtt B, Larsen FK et al (2001) Chem Eur J 7:3756–3767
31. Gatti C, Cargnoni F, Bertini L (2003) J Comput Chem 24:422–436

32. Madsen AØ, Sørensen HO, Flensburg C et al (2004) Acta Crystallogr Sect A 60:550–561
33. Hoser AA, Dominiak PM, Wozniak K (2009) Acta Crystallogr Sect A 65:300–311
34. Schomaker V, Trueblood K (1968) Acta Crystallogr Sect B 24:63
35. Madsen AØ (2006) J Appl Cryst 39:757–758
36. Munshi P, Madsen AØ, Spackman MA et al (2008) Acta Crystallogr Sect A 64:465–475
37. Koritsanszky TS (2006) Topology of X-ray charge density of hydrogen bonds. In: Grabowsky SJ (ed) Hydrogen bonding – new insights. Springer, The Netherlands
38. Desiraju GR (2005) Chem Commun 2995–3001
39. Gatti C, May E, Destro R et al (2002) J Phys Chem A 106:2707–2720
40. Gu Y, Kar T, Scheiner S (1999) J Am Chem Soc 121:9411–9422
41. Lo Presti L, Soave R, Destro R (2006) J Phys Chem B 110:6405–6414
42. Roversi P, Destro R (2004) Chem Phys Lett 386:472–478
43. Morokuma K (1971) J Chem Phys 55:1236–1244
44. Ziegler T, Rauk A (1977) Theor Chim Acta 46:1–10
45. Spackman MA (2006) Chem Phys Lett 418:158–162
46. Volkov A, Macchi P, Farrugia LJ, Gatti C, Mallinson P, Richter T, Koritsanszky TXD (2006) A Computer Program Package for Multipole Refinement, Topological Analysis of Charge Densities and Evaluation of Intermolecular Energies from Experimental and Theoretical Structure Factors
47. Farrugia LJ, Kocovsky P, Senn HM et al (2009) Acta Crystallogr Sect B 65:757–769
48. Munshi P, Thakur TS, Guru Row TN et al (2006) Acta Crystallogr Sect B 62:118–127
49. Nelyubina YV, Troyanov SI, Antipin MY et al (2009) J Phys Chem A 113:5151–5156
50. Grabowsky S, Hesse MF, Paulmann C et al (2009) Inorg Chem 48:4384–4393
51. Jayatilaka D, Grimwood DJ (2001) Acta Crystallogr Sect A 57:76–86
52. Madsen GKH, Iversen BB, Larsen FK et al (1998) J Am Chem Soc 120:10040–10045
53. Herbstein FH, Iversen BB, Kapon M et al (1999) Acta Crystallogr Sect B 55:767–787
54. Kumar GA, McAllister MA (1998) J Am Chem Soc 120:3159–3168
55. Madsen GKH, McIntyre GL, Schiøtt B et al (2007) Chem Eur J 13:5539–5547
56. Hirshfeld FL (1976) Acta Crystallogr Sect A 32:239–244
57. Madsen GKH, Wilson C, Nymand TM et al (1999) J Phys Chem 103:8684–8690
58. Piccoli PMB, Koetzle TF, Schultz AJ et al (2008) J Phys Chem A 112:6667–6677
59. Sørensen J, Clausen HF, Poulsen RD et al (2007) J Phys Chem A 111:345–351
60. Schmidtmann M, Farrugia LJ, Middlemiss DS et al (2009) J Phys Chem A 113:13985–13997
61. Gilli G, Gilli P (2000) J Mol Struct 552:1–15
62. Braga D, Grepioni F, Novoa JJ (1998) Chem Commun 1959–1960
63. Steiner T (1999) Chem Commun 2229–2300
64. Macchi P, Iversen BB, Sironi A et al (2000) Angew Chem Int Ed 39:2719–2722

Struct Bond (2012) 146: 75–100
DOI: 10.1007/430_2011_71
© Springer-Verlag Berlin Heidelberg 2012
Published online: 3 March 2012

Some Main Group Chemical Perceptions in the Light of Experimental Charge Density Investigations

U. Flierler and D. Stalke

Abstract The focus of this chapter lies on the deduction of chemical properties from charge density studies in some interesting, mainly main group element compounds. The relationship between these numerical data and commonly accepted simple chemical concepts is unfortunately not always straightforward, and often, the researcher relies on heuristic connections, rather than rigorously defined ones. In this chapter, we demonstrate how charge density analyses can shed light on aspects of chemical *bonding* and the chemical *reactivity* resulting from the determined bonding situation.

Keywords Double Bond · Experimental Charge Density · Hypervalency · Lithium · Main Group Chemistry · Reactivity · Sulfur

Contents

U. Flierler and D. Stalke (✉)
Georg-August-Universität Göttingen, Institut für Anorganische Chemie, Tammannstraße 4, 37077 Göttingen, Germany
e-mail: dstalke@chemie.uni-goettingen.de

1 Introduction

The motivation to undertake electron density studies is diverse: Besides fundamental approaches of improving the method, a deeper understanding of chemical bonding and chemical reactivity is the main concern of synthetic chemists taking advantage of this method. Experimental and theoretical electron density studies provide an ideal tool to understand the bonding in various compounds. To be able to get data of the high quality needed for an electron density study, crystals of superb quality have to be at hand. In ideal cases, experimental results can be compared to theoretical calculations. This provides additional confidence in the method in general and in the obtained results in detail. Both electron densities can be examined following *Bader's* quantum theory of atoms in molecules [1]. This approach provides a large number of descriptors that can be employed to distinguish between different bonding modes [2–8]. Compounds with controversial questions related to bonding and interesting reactivity are therefore often targets of electron density investigations. A thorough understanding is the prerequisite of a feasible deduction of the underlying principles. And once the bonding is understood, the reactivity can be estimated and modified. That way, new hidden synthetic routes might be discovered.

The focus of this chapter lies on the deduction of chemical properties from charge density studies of ten representative, mainly main group element compounds. The relationship between these numerical data and commonly accepted simple chemical concepts is unfortunately not always straightforward, and often, the researcher relies on heuristic connections, rather than rigorously defined ones. In this chapter, we demonstrate how charge density analyses can shed light on aspects of chemical *bonding* and the chemical *reactivity* resulting from the determined bonding situation.

1.1 Valence Expansion

The octet rule, as proposed by *Lewis* in 1916 [9], is still one of the fundamental concepts for discussing the electronic structure of molecules. To fulfill this rule, the valence shell of each atom in a molecule has to be filled with (exactly) eight electrons – or four (bonding or nonbonding) electron pairs. If the valence shell of the central atom in a molecule contains more or less than eight electrons, the molecule is called hyper- or hypovalent, respectively. Well-known examples for these exceptions are molecules such as BF_3, PCl_5, and SF_6.

The bonding situation in hypervalent molecules [10–12] is still a topic of constant debate. Different approaches for interpreting the bonding in these kinds of molecules have been proposed. Hybrid orbitals of the $d^n sp^3$-type pursuant to the sp^n hybridization in molecules obeying the octet rule were consulted for an explanation of the bonding situation within the valence bond (VB) theory. This description was mainly adopted for numerous textbooks even if theoretical calculations soon indicated that the d-orbitals merely serve as polarization functions and do not contribute to the bonding [13]. Closer inspection of hypervalent molecules revealed that in most of them the substituents are more electronegative than the central atom. For this reason, the compounds were now described by resonance structures avoiding hypervalency by involving the formulation of ionic bonds. However, as these structures are intentionally written to be in accordance with the octet rule, they do not prove that the octet rule is obeyed.

Another way of overcoming the dilemma of hypervalent molecules has been suggested by *Rundle* [14]. The geometry in SO_x structures makes the formation of a π-electron system feasible. The resulting m-center-n-electron p_π–p_π bonding reduces the number of valence electrons around the central atom.

However, one has to be aware that all these approaches are merely ways of describing an observed bonding situation. None of them can possibly account for the real bonding situation. As electron density studies are able to shed light on the true nature of chemical bonds, they are definitely an appropriate tool to decide which of these interpretations of the hypervalent molecule is closest to be correct.

2 Hypervalency in Silicon-Containing Compounds [15]

A potentially hypervalent atom can be found in difluoro-bis-[*N*-(dimethylamino)-phenylacetimidato-*N*,*O*] silicon (**1**). In this hexacoordinate complex (Fig. 1), the silicon atom is coordinated by two nitrogen, oxygen, and fluorine atoms, respectively [16]. Due to the differences in electronegativity, three different sets of highly polar silicon–element bonds (Si–E, E = N, O, F) are present in a molecule with one central, formally hypervalent atom. An inspection of the true nature of the silicon–element bonds will allow us to decide whether the hypervalent description of the molecule is valid.

Early studies by *Pauling* [17] already suggested that, based on the electronegativities, an Si–O bond has about 50% covalent character rather than being purely ionic. However, further studies on the still controversial nature of Si–O bonds [18] concentrated mainly on silicates. Other silicon–element bonds have barely been studied, with the exception of a theoretical study on the nature of the Si–N bond [19] and an experimental charge density study on K_2SiF_6 without a quantitative topological analysis [20].

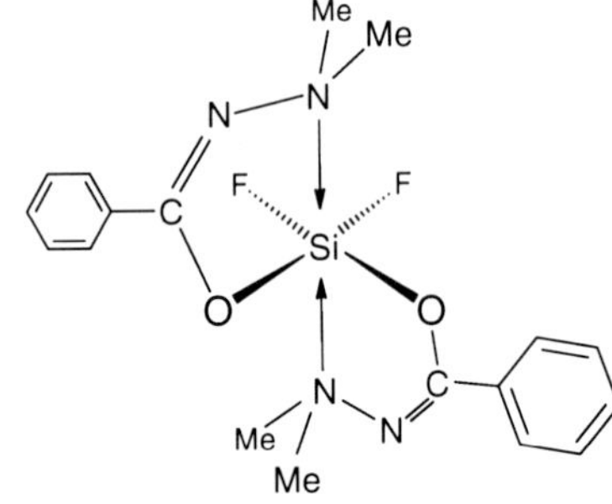

Fig. 1 Canonical formula of **1**

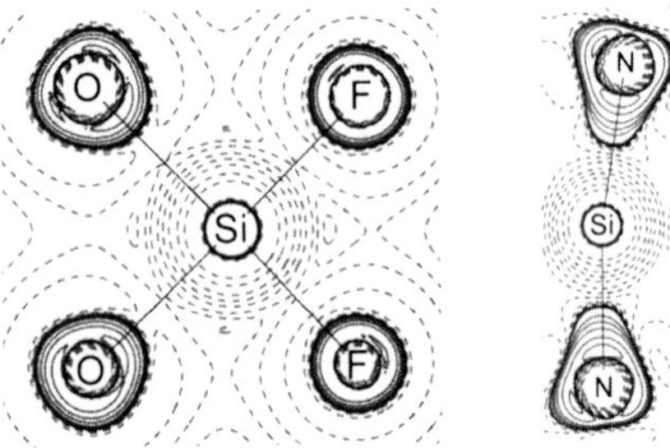

Fig. 2 Contour plot of the *Laplacian* distribution in the O_2SiF_2 (*left*) and the SiN_2 plane (*right*) of **1**. Positive values of $\nabla^2\rho(\mathbf{r})$ are depicted by *red*, and negative by *blue lines*

The topological analysis of difluoro-bis-[*N*-(dimethylamino)phenylacetimidato-*N*,*O*] silicon (**1**) was finally able to shed light on the bonding situation. The two-dimensional *Laplacian* distribution (Fig. 2) in the O_2SiF_2- and the SiN_2-plane of **1** reveals the nature of the bonding in the Si–E bonds. The spatial distribution of the *Laplacian* is almost spherical around the silicon atom, which is thus interpreted as electronically depleted. Additionally, it still shows a rather ionic-like behavior around the oxygen and fluorine atoms, indicating predominantly ionic Si–E bonds with differing – but always small – covalent bonding contributions.

The deformations of the *Laplacian* around the substituents can be directly related to their electronegativity (χ) and thus to the degree of their respective bond polarization – the lower the χ of the bonding partner, the more the deformation is pronounced. These deformations are only marginal with the oxygen atoms, while with fluorine atoms, they are virtually undetectable. The significant polarization of the density at the nitrogen atoms towards the depleted silicon atom results from the lone pairs in the apical positions at the three-coordinate amine functions.

The integrated charges of the atoms involved encourage the conclusion of the dominating ionic character of the bonds. A distinct positive charge of +2.78 e is found at the silicon atom, and negative charges are found for the fluorine (−0.80 e), the oxygen (−1.21 e), and the nitrogen (−0.78 e) atoms.

Therefore, the experimental electron density study was able to reveal the predominant ionic bonding contribution to all Si–E bonds. Thus, the hypervalent description of the silicon atom can be ruled out due to a charge transfer from the silicon to its more electronegative bonding partners. This results in a valence shell of the silicon atom with eight (or even less) electrons and a *Lewis* formula that does not violate the octet rule. However, the nature of the silicon–element bonds in these high-coordinate species is still under debate [21–23].

3 Phosphorus-Based *Janus* Head Ligands

Phosphorus-based ligand systems with one or more donating atoms in the periphery are gaining increasing importance in catalysis and in the design of self-assembling ligands. The incorporation of heteroaromatic substituents at the phosphorus atom instead of the commonly employed phenyl groups alters and augments the reactivity and coordination capability of the ligand system and leads to the design of multidentate *Janus* head ligands (for review see: [24, 25]). These ligands have the advantage of providing coordination sites for both hard and soft metal centers in terms of the hard–soft acid–base concept of *Pearson* [26]. Ligands of this type find many applications in alkane dehydrogenation, cyclometalation, cross coupling, organic transformation, catalyst recovery, antimicrobial agents, light-emitting diodes, and other electrophosphorescent devices.

3.1 *Dibenzothiazolylphosphane*

Di(benzothiazol-2-yl)phosphane, $HP(bth)_2$ (**2**), is a selected example of a *Janus* head ligand [27, 28]; rather than merely providing bulk or stereo information like a classical phosphane ligand, this heteroaromatic-substituted ligand supplies a second and even a third coordination site in addition to the divalent phosphorus(III) atom because the benzothiazolyl substituent (bth = C_7H_4NS) features a sulfur and a nitrogen atom as potential donor atoms. These substituents enable the secondary phosphanes to show a rich coordination-site selectivity towards various metal fragments.

Interestingly, in $HP(bth)_2$ (**2**), the hydrogen atom of the secondary phosphane is bonded to one ring nitrogen atom, leaving the phosphorus(III) atom divalent. This can already be seen from the NMR data from solutions of **2** and is verified by the solid-state structure from X-ray diffraction. This fact differentiates $HP(bth)_2$ from pyridyl-substituted *Janus* head ligands like $HP(py)_2$ [29], where the hydrogen atom is bonded to the phosphorus atom. So far, this alternative bonding situation has only been observed in diacylphosphanes [30, 31], which show keto-enol tautomerism in solution. Therefore, di(benzothiazol-2-yl)phosphane (**2**) could analogously be described as the N–H tautomer of the P–H form generally anticipated for secondary phosphanes.

Further NMR studies show that the N–H tautomer of **2** is stable in diethyl ether while it converts to the P–H tautomer in tetrahydrofuran. Surprisingly, the P–H tautomer is not stable, and a transformation to tris(benzothiazol-2-yl)phosphane and the primary (benzothiazol-2-yl)phosphane occurs. A comparison of the total

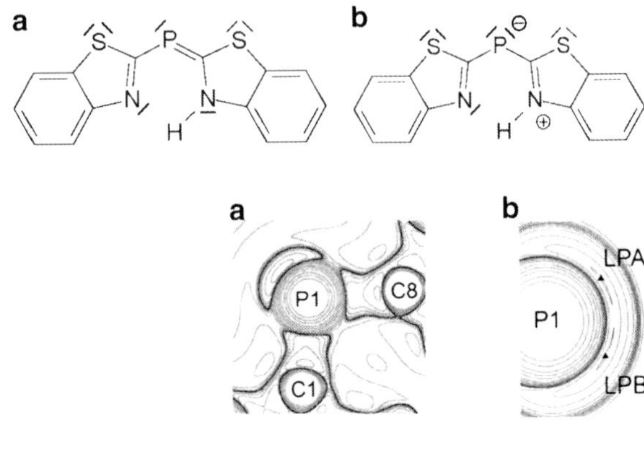

Fig. 3 Two canonical forms of $HP(bth)_2$ (**2**)

Fig. 4 Contour plot of the theoretically obtained *Laplacian* function of **2** in the plain (**a**) and orthogonal to the mean plane (**b**) of the phosphane, calculated with AIM2000. Charge concentrations (*blue lines*) refer to negative values of $\nabla^2\rho(\mathbf{r})$, and charge depletions (*red lines*) to positive values of $\nabla^2\rho(\mathbf{r})$

energies of the tautomers derived from theoretical computations indicates that the N–H tautomer is about 22 kJ/mol more stable than the best P–H tautomer.

Due to the broad applicability of *Janus* head ligands of this type, an understanding of its reactivity is vital. The central question is whether such an unusual divalent phosphorus(III) atom should be regarded a potential 2-electron (Fig. 3a) or 4-electron donor (Fig. 3b). Form **a** features an sp^2 hybridized phosphorus atom, while in form **b**, the phosphorus atom is sp^3 hybridized.

This differentiation can be made on the basis of *Bader's* quantum theory of atoms in molecules. Within this theory, (3,–3) critical points in the *Laplacian* distribution $\nabla^2\rho$, the so-called valence shell charge concentrations (VSCCs), are interpreted as bonding and nonbonding electron pairs. This way, even lone pairs in nonbonding regions can be detected that are not straightforwardly available from simple geometry considerations of routine X-ray structures.

The inspection of the *Laplacian* distribution in $HP(bth)_2$ (**2**) from DFT-based geometry optimizations shows four (3, –3) critical points around the phosphorus atom. Therefore, the phosphorus atom has to be regarded as sp^3 hybridized with two 2e2c bonds and two lone pairs (Fig. 4). This hybridization clearly supports the formulation of **2** as a 4-electron donor as described by form **b** in Fig. 3.

The detected hybridization is confirmed by the chemical reactivity of the complex. The deprotonation of $HP(bth)_2$ (**2**) with ^{n}BuLi affords the lithium phosphanide $[(Et_2O)_2Li(bth)_2P]$, in which the hydrogen atom is replaced by lithium. This metal atom is then coordinated to both nitrogen atoms of the ligand. Thus, the phosphorus atom remains two-coordinated, favoring neither of the two canonical forms over the other. However, by reacting this lithium phosphanide with $[CpMn(CO)_2(THF)]$, the lone-pair density at the phosphorus atom should be exploited to form a dinuclear organometallic complex. As expected, the THF molecules are easily replaced by other *Lewis* bases, and the remaining $[CpMn(CO)_2]$ moiety is soft enough to suit the lithium phosphanide. Therefore, in the resulting product, the hard nitrogen face of the *Janus* head ligand coordinates the lithium cation like it does in the starting material. In contrast, the soft phosphorus atom μ-bridges two $[CpMn(CO)_2]$ moieties, mimicking a 4-electron donor. The manganese atoms are located almost in the position where the lone pairs of a sp^3 hybridized phosphorus atom are anticipated. Therefore, the hybridization of the product is in accordance with the results of the theoretical calculations discussed for the precursor molecule. The QTAIM analysis was therefore able to predict the chemical behavior of this *Janus* head ligand.

3.2 *Dipyridylphosphane*

An unusual divalent phosphorus(III) atom is also present in $[Me_2Al(\mu\text{-py})_2P]$ (**3**). In this complex, the $[Me_2Al]^+$ moiety is coordinated to both pyridyl ring nitrogen atoms of the $[(py)_2P]^-$ anion [32, 33].

The results of a theoretical charge density investigation of $HP(bth)_2$ (**2**) raised the question of whether this divalent phosphorus atom should be described as a 4-electron donor or if the description as a 2-electron donor suits the molecule better (Fig. 5). Again, the hybridization of the phosphorus atom differentiates the two canonical forms. The P–C single bond would require an sp^3 hybridized phosphorus atom while an sp^2 atom would cause a conjugated P=C double bond.

In addition to the theoretical calculation as shown for **2**, compound **3** was also investigated by an experimental X-ray diffraction experiment. Inspection of the *Laplacian* at the P–C bond critical point (BCP) shows slight quantitative differences between theory and experiment; however, the contour plots show an overall similarity to each other [34]. Concordantly, both results indicate no pronounced double-bonding character of the bonds inspected (Fig. 6a, b). This is a first hint against conjugation and thus against a distinct contribution of form **a**, despite the short P–C bond path of 1.79 Å, which might erroneously be taken as an indicator for P=C double-bond character (P–C in phosphabenzene is ca. 1.74 Å and ca. 1.79 Å in phospholides).

Additionally, the positive *Laplacian* at the Al–N BCPs indicates their ionic-bonding character which is underlined by the QTAIM charge separation (exp./*theo.*: −1.21/*−1.11* and −1.15/*−1.09* e for the two nitrogen atoms and +2.04/*+3.30* e for the aluminum atom).

The orientation of the VSCCs in the nonbonding region finally confirms the distinct contribution of form **b** (Fig. 5). The VSCCs' orientation corresponds to

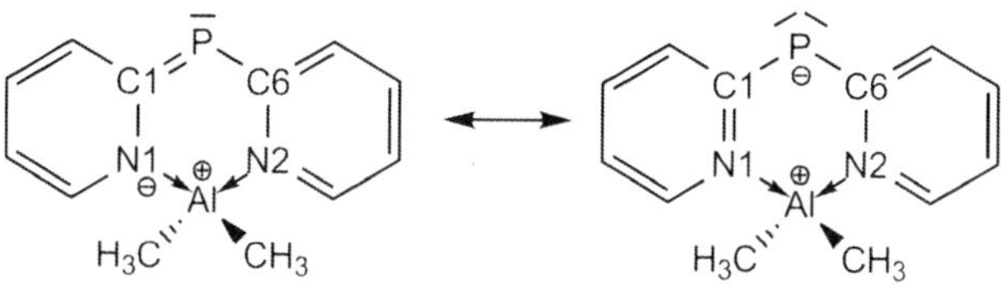

Fig. 5 Two canonical forms of $[Me_2Al(\mu\text{-py})_2P]$, **3**

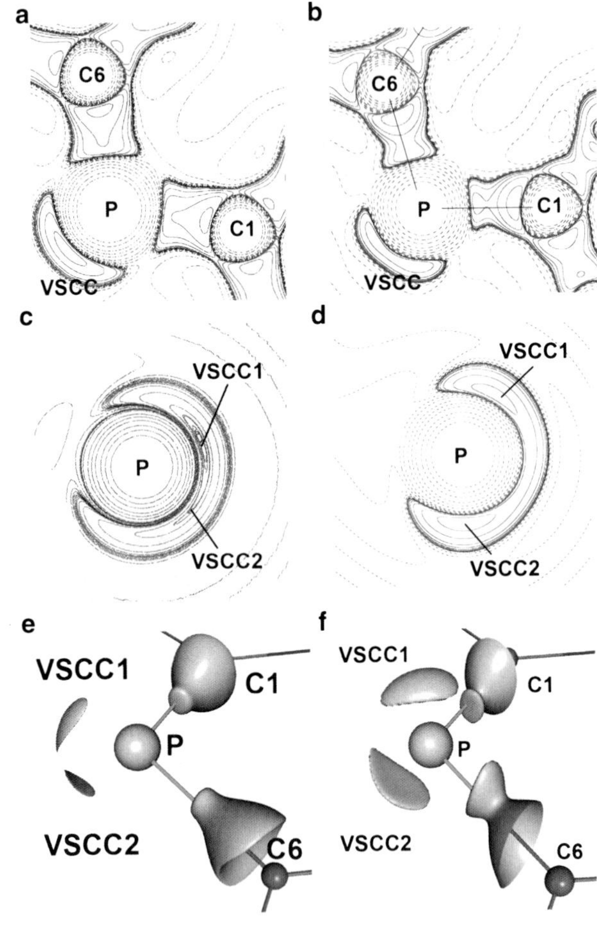

Fig. 6 Theoretically (*left*) and experimentally (*right*) obtained distributions of $\nabla^2\rho$ in the C1–P–C6 plane (**a** and **b**) and in the plane defined by *P* and the two nonbonding VSCCs (**c** and **d**) in $[Me_2Al(\mu\text{-py})_2P]$, **3**. Charge concentrations (*blue lines*) refer to negative values of $\nabla^2\rho(\mathbf{r})$, and charge depletions (*red lines*) to positive values; isosurface representation of $\nabla^2\rho(\mathbf{r})$ around P1 at the -4.9 eÅ^{-5} (**e**) and -4.0 eÅ^{-5} (**f**) level, indicating the two lone pairs in the nonbonding region

distorted phosphorus sp^3 orbitals, one VSCC above and one below the molecular plane close to the phosphorus atom (see Fig. 6e, f).

The orientation of the phosphorus atom's lone pairs seems to be suitable for accepting two $[(L_n)M]$ *Lewis* acidic organometallic residues. To test the Lewis basicity of **3** synthetically, the preparation of a dinuclear organometallic metallaphosphane complex was carried out using $[W(CO)_5(THF)]$ as a starting material.

Surprisingly, the X-ray structure analysis revealed their composition to be $[\{(OC)_5W\}_2(\mu\text{-P})(py)_2(H)]$. In the course of the reaction, the $[Me_2Al]^+$ moiety was lost and replaced by a proton to generate the μ-bridging $P(py)_2(H)$ phosphane, akin to an *N*-protonated phosphanide. In $[\{(OC)_5W\}_2(\mu\text{-P})(py)_2(H)]$, the hydrogen atom of the secondary phosphane is bonded to one ring nitrogen atom. This seems surprising as in the parent dipyridylphosphane $HP(py)_2$ the hydrogen atom is bonded to the phosphorus atom. The symmetrical coordination of the two tungsten atoms to the central phosphorus atom is in geometrical accordance with the two lone pairs of form **b** in Fig. 5. The charge density in the two lone pairs is well suited to accommodate two unsupported $W(CO)_5$ residues. The question to what extent this lone pair density is able to act as a *full* 4-electron donor, however, remains open. Even if the reactivity turned out to be not exactly predictable, the coordinating behavior could roughly be envisaged.

4 Phosphorus(V)-Based Ligands

Other interesting classes of phosphorus-based ligands are phosphonium ylides and iminophosphoranes. As isoelectronic analogues of phosphane oxides, their properties have been widely exploited in organometallic and organic synthesis [35] as well as in materials [36]. The (stereo)selective transformation of ketones and aldehydes to olefins via the *Wittig* reaction and its extensions is only one, though one very well-known, example for their broad applicability.

4.1 Iminophosphoranes

Polyphosphazenes and iminophosphoranes exhibit a thermally robust bond between the phosphorus and the nitrogen atoms. In contrast, the Si–N bond in silylated iminophosphoranes of the general type $R_3P{=}NSiMe_3$ [37] can easily be cleaved in reactions with polar organometallics in polar solvents. Because of the chemical importance of ylides, iminophosphoranes, and phosphane oxides [38], it is not unexpected that the nature of P=E bonding (E = C, N, O, S, Se) is an issue of experimental and theoretical debate.

Together with the stability, the electric conductivity of polyphosphazenes and iminophosphoranes might lead to a description of P–N bonds as double bonds, as it is often done in textbooks. To be precise, the P=N bonds in iminophosphoranes are mostly described as a resonance hybrid between a double-bonded ylenic $R_3P{=}NR$ and a dipolar ylidic form $R_3P^{\delta+}{-}N^{\delta-}R$ [35, 39, 40]. Therefore, iminophosphoranes are said to represent a thermodynamic sink, which is why attempts to reduce P^{V} to P^{III} have not been undertaken synthetically to date.

As already described above, it has been shown that *d*-orbitals in main group hypervalent compounds do not play a significant role in bonding. It has been proposed instead that negative hyperconjunction may be responsible for any *p*-character in the P–O, P–C, or P–N bonds [13, 38, 41–43]. These results have been substantiated by calculations dealing with the *Wittig*-type reactivity of phosphorus ylides [44–47] and iminophosphoranes [48–50].

The geometrical features of the anionic ligands in the metalated complex $[(Et_2O)Li\{Ph_2P(CHPy)(NSiMe_3)\}]$ (**4**) suggest canonical formulas as depicted in Fig. 7 [51]. However, all of them require valence expansion at the phosphorus atom, at

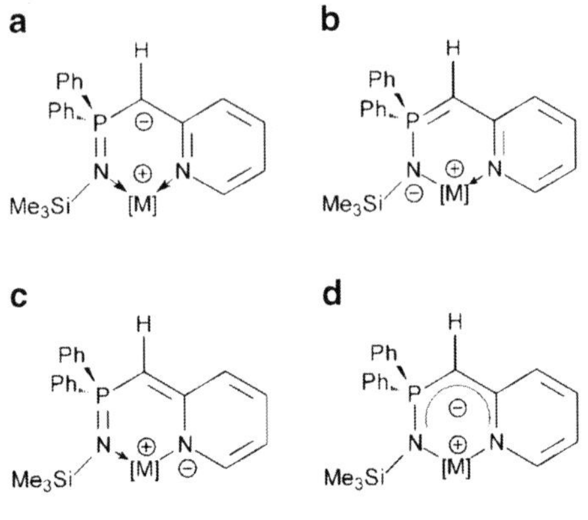

Fig. 7 Resonance forms of $[M\{Ph_2P(CHpy)(NSiMe_3)\}]$ (**4**): (**a**) indicates a carbanionic ylidic contribution, (**b**) shows the amidic ylenic resonance structure, (**c**) emphasizes the amidic olefinic resonance form, and (**d**) visualizes the delocalization of the negative charge

variance with the eight-electron rule, but chemical reactivity supports neither P=N nor P=C double bonds because both are easily cleaved in various reactions [52, 53].

An electron density investigation of this complex elucidated the bonding. The *Laplacian* distribution $\nabla^2\rho(\mathbf{r})$ along the P–C and the P–N bond paths was inspected (Fig. 8). All investigated P–C bond paths exhibit an almost equal charge concentration in the phosphorus basin. The differences in the basins of the carbon atoms stem mainly from the shorter distance between C1 and $BCP_{P\text{–}C1}$ in comparison to the other P–C bonds. The origin for this bond shortening can be seen in distinct electrostatic interactions between the negatively polarized C1 and the electropositive phosphorus atom which show charges of −0.52 e and +2.20 e, respectively.

In contrast, the *Laplacian* distribution $\nabla^2\rho(\mathbf{r})$ along the P–N1 bond shows a totally different appearance. At the BCP, it is positive and the charge density is exclusively concentrated in the nitrogen basin. This indicates a strong contribution of electrostatic interaction to the bonding energy, further substantiated by the charge integration over the atomic basins. Both nitrogen basins have negative values, however on different scales. The imino nitrogen atom, which is bonded to three electropositive neighbors, has a charge of −1.91 e and is therefore more negative than that of −1.11 e for the ring nitrogen atom. All these findings support an ylidic $P^{\delta+}$–$C^{\delta-}$ concurrent with a $P^{\delta+}$–$N^{\delta-}$ bond not present in the resonance forms of Fig. 7.

Thus, the experimental charge density investigation demonstrates that the formal P=N imino double bond and the potential ylenic P=C double bond must be written as polar $P^{\delta+}$–$N^{\delta-}$ and $P^{\delta+}$–$C^{\delta-}$ single bonds that are augmented by electrostatic contributions. This description corresponds best with the reactivity: metal organyls in polar solvents can more easily cleave this $P^{\delta+}$–$N^{\delta-}$ bond rather than the wrongly assigned P=N double bond. Therefore, deimination or the retro-Staudinger reaction of iminophosphoranes seems an unorthodox but suitable synthetic route to phosphanes. With this knowledge, P^V–N molecules can be reduced with polar metallorganic bases in polar solvents. In this way, access to chiral phosphane amines was opened up, which are difficult to synthesize by any other method. Thus, an incorrect perception of the P=N double bond had blocked the synthetic access to these compounds for many years [52, 53].

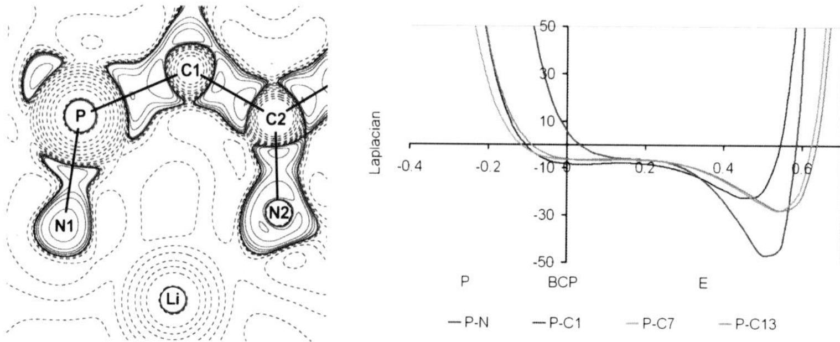

Fig. 8 Contour representation of $\nabla^2\rho(\mathbf{r})$ in the LiN_2C_2P-plane (*left*) of [$(Et_2O)Li\{Ph_2P(CHPy)(NSiMe_3)\}$] (**4**) and the $\nabla^2\rho(\mathbf{r})$ distribution along the bond paths (*right*)

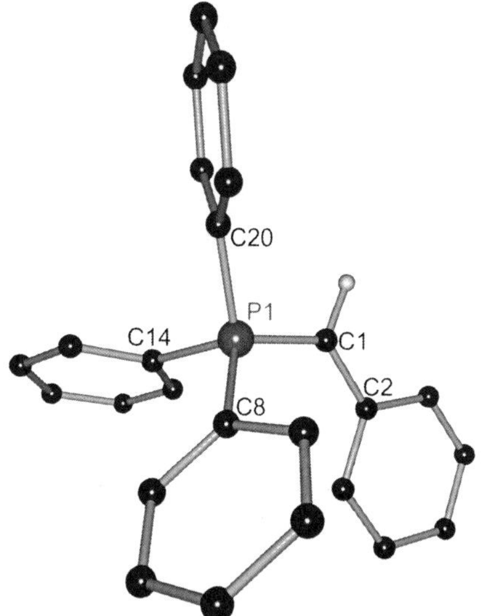

Fig. 9 Molecular graph of the triphenylphosphonium benzylide Ph_3PCHPh (**5**)

4.2 *Phosphorus Ylide [54]*

Another interesting class of phosphorus-based compounds is that of phosphorus ylides. They play a crucial role in synthetic organic chemistry [55]. The nature of the P–C_{ylide} bond in these compounds therefore is in the focus of research, and a number of studies based on theoretical calculations at different levels have been performed. However, there are very few examples of experimental studies on the electronic distribution in hypervalent phosphorus compounds and in ylides in particular. The investigation of the semistabilized ylide, triphenylphosphonium benzylide Ph_3PCHPh (**5**) (see Fig. 9), is one rare example.

The topological analysis of the experimentally determined charge density distribution in **5** revealed the multiple-bond character and high ionicity of the ylide bond.

The *Laplacian* values at the BCPs of the P–C bonds are negative, indicating shared interactions and thus covalent bonding. Even if the *Laplacian* was not

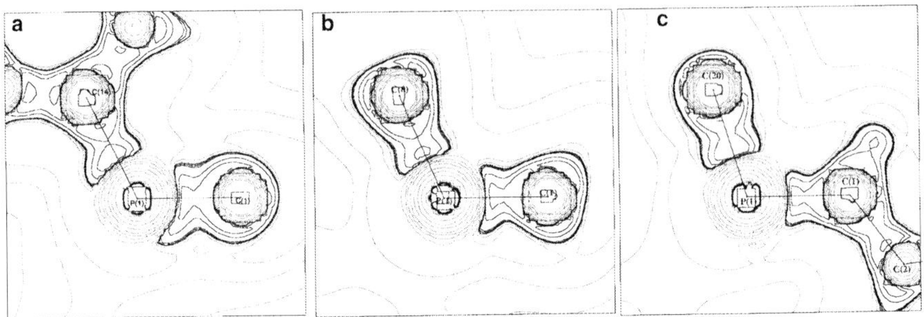

Fig. 10 The *Laplacian* maps of **5** in the planes C(1)P(1)C(14) (**a**), C(1)P(1)C(8) (**b**), and C(1)P(1) C(20) (**c**). Contours at logarithmic intervals in $-\nabla^2\rho/e\text{Å}^{-5}$

investigated along the whole bond path, these results are in accordance with the previously presented results.

On the other hand, the ylide bond proves to be very polar: net atomic charges for P and C(1) are +0.49 e and −0.41 e, respectively. In addition, a high ellipticity of the ylide bond P–C(1) is observed. This implies the multiple character of these bonds and a high degree of delocalization over the P–C(1)–Ph system. The *Laplacian* distribution of the P–C(1) and P–C(14) bonds is distinctively asymmetric (Fig. 10), with a shift of the charge concentration towards the area of σ^*-orbitals of the P–C bonds. This is not the case for the other two P–Ph bonds (Fig. 10).

The positions of the BCPs relative to the atoms reveal further interesting information. In the case of different atomic charges of the covalently bonded atoms, one can expect to observe a shift of the BCP towards the more positive atom, which reflects the accumulation of electron density in the atomic basin of the more negative atom. This is observed for the P–C(1) and P–C(8) bonds. However, for the P–C(20) and P–C(14) bonds, as well as for the C(1)–C(2) bond, a shift in the opposite direction is found. It seems that these shifts of the BCP positions reflect the charge transfer from the stabilizing phenyl group towards the (PPh_3) fragment and show the different roles played by the Ph rings in delocalization of the charge. High values of the relative shift indicate that the shape of the atomic basin of the phosphorus deforms quite easily.

The topological analysis of the experimentally determined charge density distribution in semistabilized ylide **4** revealed the multiple character and high ionicity of the ylide bond.

5 Sulfur-Based Ligands

5.1 Sulfur Ylide

Similar to the *Wittig* phosphonium ylides, the *Corey* sulfur ylides ($R_2S^{\delta+}–{}^{\delta-}CR_2$) are widely used in organic synthesis for stereoselective epoxidations, cyclopropane formations, and ring expansion reactions [56–64]. Nevertheless, their electronic properties are still under debate.

$[(thf)Li_2\{H_2CS(NtBu)_2\}]_2$ (**6**) is a compound of striking interest because it resembles a sulfur ylide and shows an Li_3CR motif, well known from organolithium chemistry [65]. The $[H_2CS(NtBu)_2]^{2-}$ dianion is an analogue of SO_3^{2-}, in which two oxygen atoms are isoelectronically replaced by a N*t*Bu imido group, and the third one is substituted by a CH_2 group [66–70]. The sulfur-bonded heteroatom groups form a tridentate ligand. Thus, formal hypervalency can be investigated along with the controversial interaction between a carbanion and a Li_3 triangle. Details on the bonding situation of lithium will be discussed in another chapter.

Various resonance formulas are feasible for the description of $[(thf)Li_2\{H_2CS(NtBu)_2\}]_2$ (**6**) (Fig. 11). However, the S–C as well as the S–N bond cleavages [71] clearly contradict the classical *Lewis* notation of S=C or S=N double bonds (hypervalent ylenic form, Fig. 11a, b) [72]. Thus, an ylidic resonance form seems much more reasonable (Fig. 11b–d) [73]. This fuels the debate as to what extent sulfur ylides are dominated by *ylidic* or *ylenic* bonding (Fig. 12). By analyzing the topological properties of the experimental electron density distribution, it was possible to identify four VSCCs and specify S–N and S–C bonds in **5** as classical single bonds strengthened by electrostatic interactions ($S^{\delta+}$–$N^{\delta-}$ and $S^{\delta+}$–$C^{\delta-}$) [74, 75].

Again, the hybridization of the atoms involved in the formal hypervalency sheds light on the true bonding situation. In **6**, four VSCCs are found at the sulfur atom: three of them directed towards bonded neighbors and one nonbonding charge concentration. These VSCCs include angles that are much closer to the ideal tetrahedral angles than the planar connectivity-related angles suggest. They range from 102.2° to 107.5°, compared with 100.31° to 103.97° for the equivalent values resulting from straight line atom connectivities.

Around the nitrogen atom, an additional four VSCCs are found. They are all directed towards bonding neighbors, i.e., towards S1, C1, and the two lithium atoms Li1 and Li3. The included angles again are very close to those anticipated for a tetrahedral arrangement. Thus, an sp^3 hybridization must be assumed for the nitrogen atom as well, and S=N double bonding is precluded.

A similar arrangement is found around the carbon atom. Four VSCCs are found: one is directed towards the sulfur atom, two to the hydrogen atoms, and one VSCC is directed roughly towards the midpoint of the Li_3 triangle.

So, from the charge density point of view, a hypervalent sulfur species can clearly be ruled out, and thus the resonance form **d** in Fig. 11 best describes the

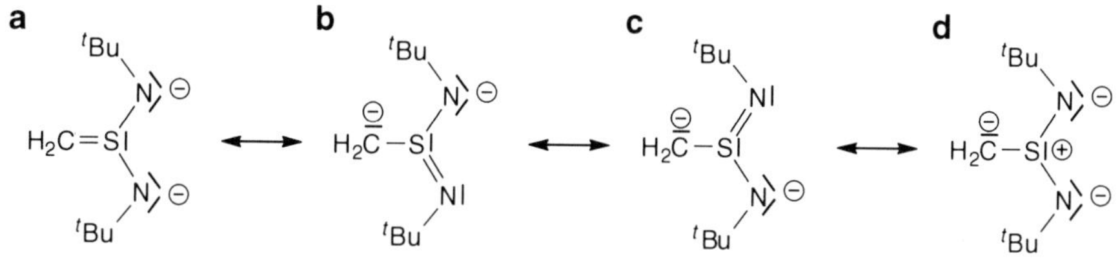

Fig. 11 Possible resonance formulas of the diimido sulfur ylide dianion, only the last (*d*) not exceeding the octet at the central sulfur atom

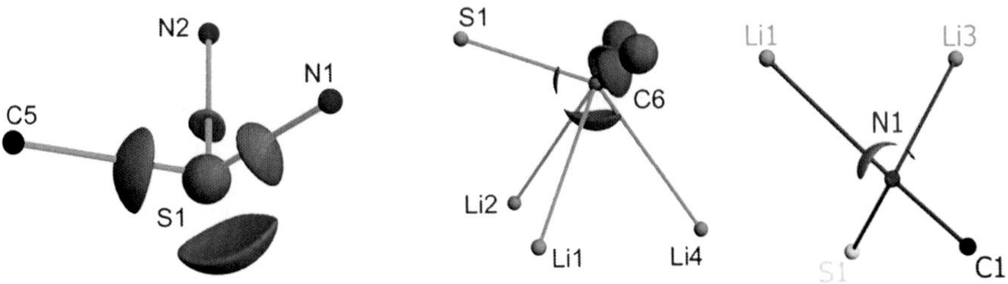

Fig. 12 Spatial distribution of the VSCCs around the sulfur atom (*left*) in $[(thf)Li_2\{H_2CS(NtBu)_2\}]_2$ (**6**) and at the carbanion, supporting the $H_2C^{\delta-}...Li_3^{\delta+}$ 4c2e bond (*middle*) and the imido nitrogen atom (*right*)

electronic features of **6**. The S–N and S–C bonds are identified as classical single bonds strengthened by electrostatic interactions ($S^{\delta+}$–$N^{\delta-}$ and $S^{\delta+}$–$C^{\delta-}$). Therefore, **6** should be formulated as *ylidic* rather than *ylenic*. This best describes the electronic situation and also explains the reactivity of the compound. Furthermore, **6** provides the first experimental evidence of a $H_2C^{\delta-}...Li_3^{\delta+}$ 4-center-2-electron bond.

5.2 *Sulfur Imides*

Widely accepted examples of formal hypervalent species were also the imide analogues of SO_2 and SO_3, the sulfur diimide $S(NtBu)_2$ (**7**) [76] and triimide $S(NtBu)_3$ (**8**) [77]. This assumption was supported by the very short distances for the sulfur–nitrogen bonds of approximately 1.5 Å, which led to the formulation of S=N double bonds in those compounds [78].

This description avoids formal charges (*Pauling's* verdict) but implies valence expansion and *d*-orbital participation in bonding at the central sulfur atom. However, this formulation is in contrast to theoretical investigations from the mid-1980s, which verified that *d*-orbitals cannot participate in the sulfur–nitrogen bonding due to large energy differences between the sulfur *p*- and *d*-orbitals [13, 38, 41, 47, 79–81]. Furthermore, these MO calculations on second-row atoms in "hypervalent" molecules showed that the *d*-orbitals are mainly needed as polarization functions rather than as bonding orbitals [82, 83]. Theoretical studies of SO_2 and SO_3 show that the S–O bonds have highly ionic character and bond orders close to one.

A different bonding mode was first suggested by *Rundle* [14, 84, 85]. He pointed out that the planarity of the SO_x units allows the formation of a delocalized π-electron system leading to *m*-center-*n*-electron bonding [86]. Several experimental observations in recent years [87–94] do not suit the idea of a classical S=N double bond either, e.g., the reactivity of many polyimido sulfur species in polar media. They easily perform transimidation reactions [92] and generate diimides [95], or the S–N bond inserts into an M–C bond [71]. Since such reactions require

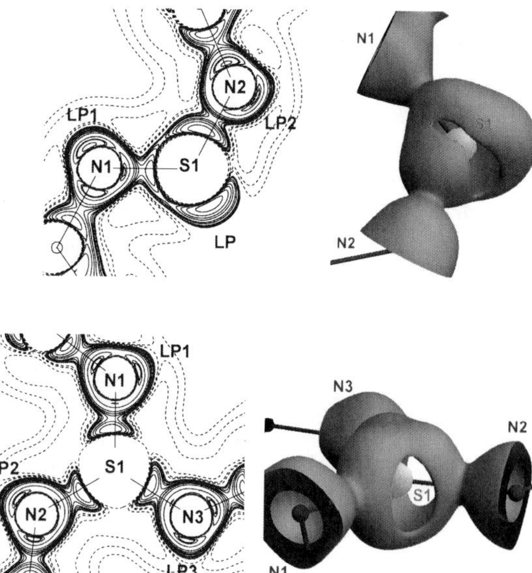

Fig. 13 contour representation of $\nabla^2\rho(\mathbf{r})$ (*left*) and reactive surface ($\nabla^2\rho(\mathbf{r}) = 0$ eÅ^{-5}) around the sulfur atom (*right*) in the sulfur diimide $S(NtBu)_2$ (**7**)

Fig. 14 Contour representation of $\nabla^2\rho(\mathbf{r})$ (*left*) and reactive surface ($\nabla^2\rho(\mathbf{r}) = 0$ eÅ^{-5}) around the sulfur atom (*right*) in the sulfur triimide $S(NtBu)_3$ (**8**)

facile S–N bond cleavage in polar media, the reactivities indicate a quite polar bonding situation rather than p_π–d_π double bonding. Furthermore, the reassignment of the SN stretching vibrations in the Raman spectrum to much lower wave numbers (640 and 920 cm^{-1} [95] instead of the previously assumed 1200 cm^{-1} [96–99]) indicates a weaker bond and probably another bonding type rather than S=N. Indeed, the S–N bonds in both compounds **7** and **8** were found to be polar in the topological analysis of the experimentally derived electron density distributions [100].

Experimental as well as theoretical *Laplacian* distributions in both planar $S(NtBu)_2$ (**7**) and $S(NtBu)_3$ (**8**) molecules reveal one single in-plane lone-pair VSCC at all nitrogen atoms. At the sulfur atom primarily sp^2 hybridization is indicated by the in-plan-coordinated substituents (Figs. 13 and 14).

The sp^2 hybridization is indicative of a π-system above and below the SN_x plane. Such a π-system is also reflected in the corresponding π-orbitals and the leading resonance structures given by an NBO/NRT (natural bond orbital/natural resonance theory) approach. This bonding type corresponds to a 4-center-6-electron bond. As a consequence of the π-system, the redistribution of charge should be quite efficient. Indeed, the NBO/NRT analyses reveal increased covalent contributions to the S–N bond orders, accompanied by decreased charges at the nitrogen atoms in **7** and **8**, compared to molecules in which the sulfur atom is sp^3 hybridized. However, from the shape of the orbitals and from the NBO/NRT resonance structures, it is obvious that the π-orbitals are polarized. Thus, the ionic contributions to the total bond orders are significant in the short S–N bonds of **7** and **8**. Again, valence expansion at the sulfur atom can definitely be excluded.

In addition to the bonding type, the investigations were also able to shed light on the experimentally observed reactivity by inspection of the reactive surfaces ($\nabla^2\rho(\mathbf{r}) = 0$ eÅ^{-5}; Figs. 13 and 14). $S(NtBu)_3$ (**8**), for example, reacts smoothly with MeLi and PhCCLi but not with *n*BuLi or *t*BuLi. The topological analysis shows that this discrimination of large reactants can be related to small areas of strong charge depletion in the SN_3 plane at the bisectors of the N–S–N angles. The carbanionic nucleophile must approach the sulfur atom along the NSN bisector in the SN_3 plane or at an angle of less than about 45°, which is only feasible for small or planar carbanions. Bulky anions cannot reach the holes, due to the steric hindrance of the N*t*Bu groups. The steric argument would not be valid if a direct orthogonal attack above or below the SN_3 plane was favored, as there is sufficient space in the planar molecule to reach the sulfur atom directly from this direction.

Electron density studies can thus not only be applied to understanding bonding situations but also allow the deduction of chemical reactivity. What is more, they can even be applied to predict reactivities and thus suggest new synthetic reaction pathways.

As expected, the S–N bonds in all of the described examples are found to be quite polar. However, for very polar bonds, problems arise if the topology is discussed exclusively at the BCPs. The BCPs of such bonds appear in a region where the electron density distribution is very flat. As a consequence, small changes in the description of $\rho(\mathbf{r})$ already lead to large alterations in the position of the BCPs. One consequence is the considerable difference between theoretically and experimentally determined values. Good agreement between experimental and theoretical results, however, can be expected for the geometrical properties and for the qualitative features of the spatial distribution of the *Laplacian* (shape of $\nabla^2\rho(\mathbf{r})$, number, and positions of nonbonding VSCCs) [101].

6 Boron-Containing Compounds

6.1 Borylene

There are numerous compound classes in which a transition metal atom is bonded to a boron atom. The nature of the bond between these two atoms, which controls the structure and the reactivity of a compound, is manifold and in many cases not unambiguously defined [102, 103]. While *borides*, *metallaboranes*, and *transition metal complexes with boron heterocycles* have been known for quite some time, *transition metal complexes of boron* constitutes a rather new compound class [104–106]. This compound class can be subdivided into three groups depending on the coordination number of the boron atom and the number of transition metal–boron bonds, namely, borane, boryl, and borylene complexes.

Due to the structural relation to the isolobal carbonyl group, the borylene ligand has attracted special interest [105, 107–111]. Borylene ligands show, just as their carbon analogues, different coordination modes to the transition metal atom. It is observed that the transition metal–boron bond is more stable against cleavage than the corresponding transition metal–carbon bond [112–114]. This can mainly be assigned to the more advantageous energetic level of the σ donor orbitals as the energy of the π^*-orbitals stays roughly the same. The narrow HOMO–LUMO gap induces a positive charge at the boron atom, which is therefore susceptible towards nucleophilic attack. Kinetic stability can be achieved by steric shielding from bulky ligands at the boron atom. In addition, it was shown that selected metal fragments, for example, $\{CpMn(CO)_2\}_2$, result in a reduction of the kinetic instability by lowering the imbalance between the HOMO and the LUMO.

Therefore, a charge density investigation of the bridged borylene complex, $[\{Cp(CO)_2Mn\}_2B(\mu\text{-}B^tBu)]$ (**9**), was of particular interest [107, 115, 116]. The predicted kinetic instability of the borylene is mitigated in this compound by two factors: on the one hand, the complexation was achieved with the $\{CpMn(CO)_2\}_2$-fragment, which, as already described above, reduces the instability and, on the other hand, by the borylene ligand being a *tert*-butyl-borylene, which has a sterically demanding organic group, additionally shielding the boron atom against nucleophilic attack.

The coordination of the borylene ligand to the transition metal shows significant differences with respect to bridging carbonyl complexes [5]. In the latter case, two VSCCs are found at the carbonyl carbon atom, one of which points towards the carbonyl oxygen atom, while the other one is broadened and directed towards the middle of the metal–metal bond [117–119]. At the borylene boron atom, however, three VSCCs are found each pointing in the direction of one of the bonding partners (Fig. 15). The identification of the three VSCCs at the boron atom by DFT methods heavily relies on the employed functional [116]. The examination of the bond paths between the transition metal atoms and the boron atom shows a pronounced curvature, which is indicative of bond delocalization. Inspection of the angles between the bond paths and the direct atom–atom vectors at the boron atom and the two manganese atoms allows for a quantification of this delocalization. The dimension of the angles implies a dominant direct and symmetrical donating interaction from the boron to the manganese atoms and an indirect, less pronounced, and unsymmetrical manganese–boron back bond.

Besides the transition metal–boron bond in the bridging molecule, another structural motif of continuing interest was present. For bridging carbonyl complexes, it is postulated that no metal–metal bond exists [5]. This is also observed with the borylene complexes. Neither the theoretically nor the experimentally derived electron densities showed a bond path between the two manganese atoms. Lately, there has been some discussion whether the absence of a bond path implies the absence of a bond [120, 121]. Only slight variations in the geometry of semibridged iron carbonyl complexes cause the abrupt disappearance of the BP, while other bond descriptors change in a physically more meaningful way [122]. Apparently, the formation of a BP can also be averted by a general increase of the electron density level in the bonding region of three-membered rings due to the core density [123]. An examination of the total electron density $\rho(\mathbf{r})$ already reveals an indentation in the Mn–Mn region and the orthogonal gradient-vector trajectories explain why the formation of an Mn–Mn BP is

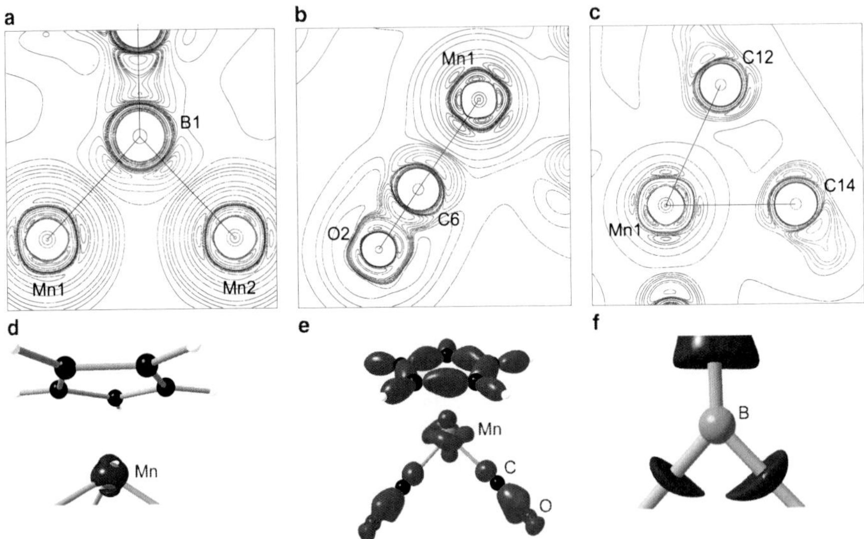

Fig. 15 $\nabla^2\rho(\mathbf{r})$ of **9** in the Mn_2B (**a**), in the Mn1–C6–O2 (**b**), and in the Mn1–C12–C14 plane (**c**) (*blue lines* indicate negative values in $\nabla^2\rho(\mathbf{r})$, while *red lines* stand for positive values in $\nabla^2\rho(\mathbf{r})$); isosurface representation of $\nabla^2\rho(\mathbf{r})$ (**d**) and static deformation density (**e**) at the Mn atoms; and isosurface representation of $\nabla^2\rho(\mathbf{r})$ at the B atom (**f**)

prevented: The atomic basin of the boron atom is spread out between the two manganese basins and inhibits the formation of a saddle point in $\rho(\mathbf{r})$ (Fig. 16).

An examination of the source function (see also the chapter by Carlo Gatti in this book) illustrates that the manganese atoms withdraw density from the bond rather than donate. This implies the presence of a nonlocalized bond. As the complex under investigation does not show any sign of paramagnetism, a superexchange of the electrons via the boron atom seems feasible, the more so as bond paths, found between the manganese atoms and the boron atom, were described to serve a privileged exchange channel for electron interactions.

Even if no bond critical point was found, an increased electron density – albeit at a very low level – was found between the two manganese atoms. This could also be interpreted as some kind of interaction. Therefore, we conclude that either more investigations or the development of more specific descriptors are necessary to describe bonding situations like metal–metal bonds unambiguously.

6.2 Electrophilic Boranes

Electrophilic boranes are – amongst other applications – used for the generation of electron-deficient early transition metal cations. Via the abstraction of an alkyl group, they activate the catalyst precursor, which can then be employed in

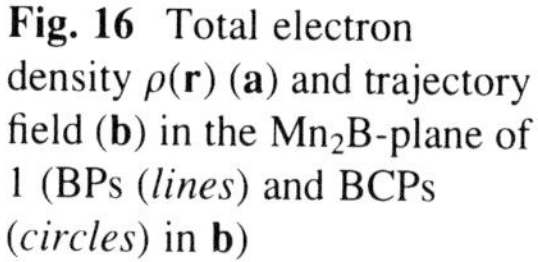

Fig. 16 Total electron density $\rho(\mathbf{r})$ (**a**) and trajectory field (**b**) in the Mn_2B-plane of 1 (BPs (*lines*) and BCPs (*circles*) in **b**)

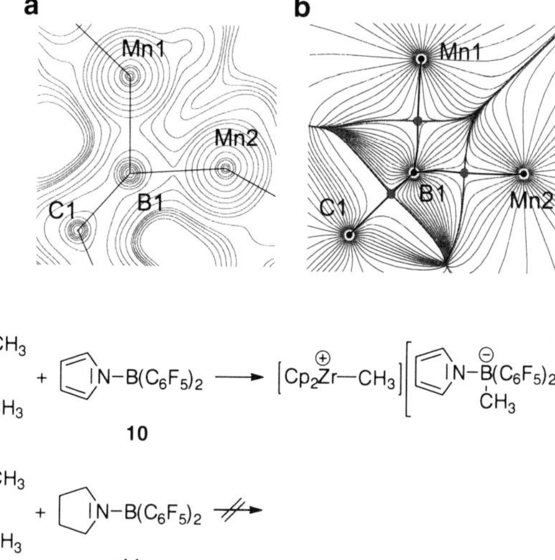

Fig. 17 Example of a reaction of **10** and **11** with a zirconocene complex

polymerization reactions [124–126]. As the cations formed often build ion pairs with the co-catalyst, i.e., the electrophilic borane, the properties of the borane influence the activity of the catalyst. More recently, they have been employed in metal-free catalysis as frustrated Lewis pairs [127].

The electrophilic tris(pentafluorophenyl)borane $B(C_6F_5)_3$ is a widely applied reagent for this kind of reaction, as the boron atom shows a high *Lewis* acidity due to the electron-withdrawing pentafluorophenyl groups. Owing to these very pronounced electron-withdrawing characteristics, a precise adjustment of the reactivity can be achieved by a variation of these ligands. The group of *Erker* succeeded in synthesizing borane complexes in which one of the pentafluorophenyl substituents is replaced by a pyrrolyl or a pyrrolidinyl ring, respectively [128]. The reactivities of the two compounds are fundamentally different: while the pyrrolyl-substituted derivative, $[(C_6F_5)_2B(NC_4H_4)]$ (**10**), still shows a marked *Lewis* acidity and thus can be utilized as a co-catalyst, the pyrrolidinyl-substituted derivative, $[(C_6F_5)_2B(NC_4H_8)]$ (**11**), no longer shows this ability (Fig. 17). As in most cases *Lewis* acidity relies upon an electronic depletion, examination of the electronic structures of the molecules allows inferences on the reasons for the change in reactivity [129].

In both molecules, the electron-withdrawing pentafluorophenyl substituents cause a pronounced electronic depletion at the boron atoms. This is clearly reflected by the positive *Bader* charges and the positive electrostatic potentials of the boron atoms (see Fig. 19). However, a difference in magnitude can be observed. While this electronic depletion is partly compensated by the pyrrolidinyl substituent, this can hardly be observed for the pyrrolyl substituent. The compensation of the electronic depletion is achieved via a delocalization of the lone pair from the nitrogen atom's p-orbital into the B–N bond or the coupling of the lone pair into the delocalized system resulting in its expansion over the B–N bond. From electron density studies, these p_π-p_π interactions can be concluded from an ellipticity along the B–N bond differing significantly from zero. Again gradual differences can be observed: the ellipticity along the B–N bond in $[(C_6F_5)_2B(NC_4H_8)]$ **11** is higher than in $[(C_6F_5)_2B(NC_4H_4)]$ **10** (Fig. 18). Consequently, the delocalization of the electron density over the B–N bond is more pronounced in **11** than in **10**.

This can easily be understood when keeping in mind that a complete coupling of the nitrogen atom's electron pair into the B–N bond would draw electrons off the aromatic system in $[(C_6F_5)_2B(NC_4H_4)]$ and thus disturb the aromaticity. Instead, the aromatic system expands out of the ring system into the B–N bond. Compared to that, the B–N bonding situation in the molecule with the nonaromatic heterocycle features a lone pair at the nitrogen atom which can totally be used for the B–N bond reinforcement. This leads to an increased π-character of the bond in **11** relative to **10**.

The different strength of the bonding of the two nitrogen atoms to their neighboring atoms is also reflected by the values of the VSCCs at the nitrogen atoms. In the pyrrolyl-substituted derivative, the VSCCs pointing in the direction of the ring carbon atoms show higher values (-73.1 eÅ^{-5}) compared to the VSCC directed towards the boron atom (-67.71 eÅ^{-5}). In the pyrrolidinyl-substituted derivative, however, the boron-directed VSCC dominates (towards B: –70.2, towards the ring carbon atoms: –58.3 and -65.2 eÅ^{-5}).

The two-dimensional distribution of $\nabla^2\rho(\mathbf{r})$ in the principal mean plane of the heterocycle substituent including the boron atom reveals the expected features of shared and – at least for B–N – polarized bonds in both molecules. Charge concentrations in the bond from both atoms form a saddle-shaped distribution with tailing perpendicular to the bonding vector (Fig. 19).

These gradual differences of the descriptors, which can solely be ascribed to the aromatic or nonaromatic character of the newly introduced substituents and the resulting differences in the nature of the B–N bond, are crucial for the reactivity.

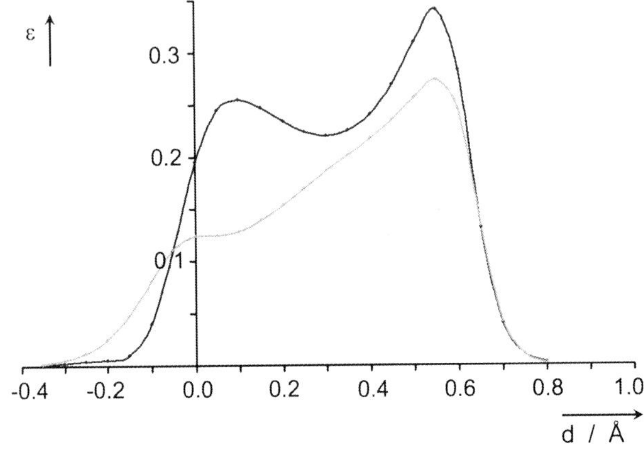

Fig. 18 Ellipticity along the B–N bond for $[(C_6F_5)_2B(NC_4H_4)]$ **10** (*dark gray*) and $[(C_6F_5)_2B(NC_4H_8)]$ **11** (*light gray*) with *d* being the distance from the BCP (at 0.0 Å) and the boron basins spanning the negative values while the nitrogen basins span the positive ones

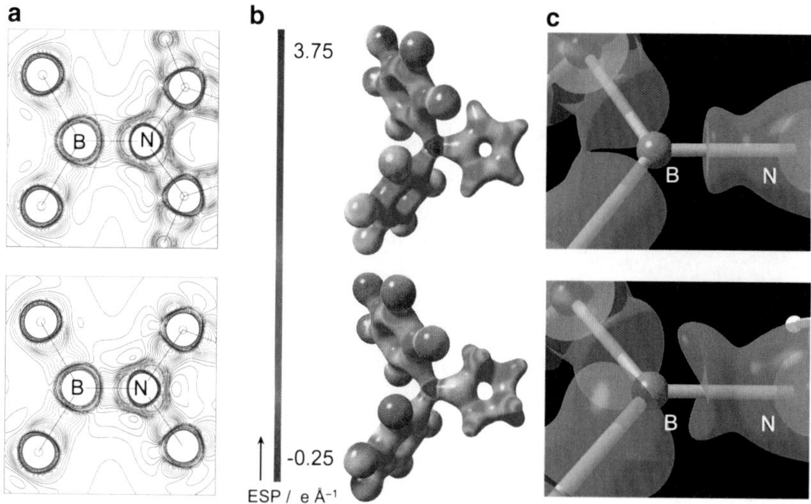

Fig. 19 Contour representation of $\nabla^2\rho(\mathbf{r})$ (**a**), isosurface representation of the electrostatic potential ranging from −0.25 to +3.75 eÅ^{-1} mapped on $\rho(\mathbf{r}) = 0.65$ eÅ^{-3} (**b**) and reactive surface ($\nabla^2\rho(\mathrm{r}) = 0$ eÅ^{-5}) (**c**) in the B–N–C_{ortho} planes of $[(C_6F_5)_2B(NC_4H_4)]$, (**10**), (*top row*) and $[(C_6F_5)_2B(NC_4H_8)]$, (**11**), (*bottom row*)

A very descriptive interpretation of the differences in the reactivity is given by the reactive surface. Holes in this isosurface of the *Laplacian* are points for nucleophilic attack, or expressed the other way round, electrophilic parts of the molecule. While the isosurface around the boron atom in $[(C_6F_5)_2B(NC_4H_4)]$ exhibits exposed areas on top and bottom, the boron atom in $[(C_6F_5)_2B(NC_4H_8)]$ is electronically shielded by claws formed by this isosurface (Fig. 19).

Thus, with the help of experimental electron density studies, the reasons for the different reactivities of $[(C_6F_5)_2B(NC_4H_4)]$ and $[(C_6F_5)_2B(NC_4H_8)]$ could be determined. This clearly demonstrates that experimental electron density studies can give definite answers to chemical questions. The examination of these two model compounds of known reactivity provides a relation between the electronic structure and the reactivity of the two compounds. An expansion of these studies to other *Lewis* acids of the same structural type will allow for a quantification of these relations. Frustrated Lewis pairs, for example, could be another landmark for which gradation in reactivity could be linked to the descriptors from electron density studies. This relation between structure and reactivity could noticeably ease the development of enhanced co-catalysts.

7 Conclusion and Outlook

The recent results from experimental charge density investigations show that the method is capable of giving answers to controversially discussed bonding issues and to provide innovative new synthetic concepts. The close linking of methodical charge density results and preparative concepts assures that the results of the first will generate a common language more widely used than currently available. The scientific synergy results from increasing the field of view: methodologists work on chemically relevant systems and synthetically oriented chemists get new concepts via the deeper insight provided for their systems. Charge density studies in molecular chemistry will transfer the intuitive-heuristic instruments, currently employed to judge reactivity (nucleophilicity, I and M effects, conjugation, delocalization, partial charges, etc.), into quantifiable figures. In the long term, they will provide the basis for the rational design of synthetic building blocks. In life sciences, drugs and cofactors are studied with respect to the charge density distribution. Deductible physical, chemical, and structure-chemical features and the molecular interaction in various crystal lattices will be related to their biological activity. In the area of materials science, only the detailed knowledge of the charge density distribution permits the de novo design and improvement of known materials of a required property profile. Almost any material specification such as magnetic, optical, and electronic performance can be deduced from the knowledge of the electronic and geometric structure.

Acknowledgement This work was supported by the Deutsche Forschungsgemeinschaft within the priority program 1178 *Experimental charge density as the key to understand chemical interactions*; the DNRF funded *Center for Materials Crystallography*; and the PhD program *CaSuS*, Catalysis for Sustainable Synthesis, funded from the Land Niedersachsen, Chemetall, Frankfurt, and the Volkswagenstiftung. The authors are particularly indebted to many capable students providing the basic results brought about by this article.

References

1. Bader RFW (1990) Atoms in molecules – a quantum theory. Oxford University Press, New York
2. Coppens P (1985) Coord Chem Rev 65:285–307
3. Koritsanszky T, Coppens P (2001) Chem Rev 101:1583–1627
4. Lecomte C, Souhassou M, Pillet S (2003) J Mol Struct 647:53–64
5. Macchi P, Sironi A (2003) Coord Chem Rev 238–239:383–412
6. Coppens P (2005) Angew Chem 117:6970–6972
7. Gatti C (2005) Z Kristallogr 220:399–457
8. Stalke D, Ott H (2008) Nachrichten aus der Chemie 56:131–135
9. Lewis GN (1916) J Am Chem Soc 38:762–785
10. Gillespie RJ, Silvi B (2002) Coord Chem Rev 233–234:53–62
11. Noury S, Silvi B (2002) Inorg Chem 41:2164–2172
12. Pierrefixe SCAH, Stralen SJMv, Stralen JNPv, Guerra CF, Bickelhaupt FM (2009) Angew Chem Int Ed 121:6501–6593
13. Kutzelnigg W (1984) Angew Chem 96:262–286
14. Rundle RE (1947) J Am Chem Soc 69:1327–1331
15. Chuit C, Corriu RJP (1998) Chemistry of hypervalent compounds, Chapter 4
16. Kocher N, Henn J, Gostevskii B, Kost D, Kalikhman I, Engels B, Stalke D (2004) J Am Chem Soc 136:5563–5568
17. Pauling L (1939) The nature of the chemical bond. Cornell University Press, Ithaka

18. Gibbs GV, Downs JW, Boisen MB Jr (1994) Rev Mineral 29:331–368
19. Wang J, Eriksson LA, Boyd RJ, Shi Z, Johnson BG (1994) J Phys Chem 98:1844–1850
20. Herster JR, Malsen EN (1995) Acta Crystallogr Sect B 51:913–920
21. Bombicz P, Kovács I, Nyulászi L, Szieberth D, Terleczky P (2010) Organometallics 29:1100–1106
22. Sidorkin VF, Doronina EP (2009) Organometallics 28:5305
23. Fester GW, Wagler J, Brendler E, Böhme U, Gerlach D, Gerlach D (2009) J Am Chem Soc 131:6855
24. Mahalakshmi L, Stalke D (2002) Struct Bonding 103:85
25. Baier F, Fei Z, Gornitzka H, Murso A, Neufeld S, Pfeiffer M, Rüdenauer I, Steiner A, Stey T, Stalke D (2002) J Organomet Chem 661:111
26. Pearson RG (1963) J Am Chem Soc 85:3533–3539
27. Stey T, Henn J, Stalke D (2007) Chem Commun 413–415
28. Stey T, Pfeiffer M, Henn J, Pandey SK, Stalke D (2007) Chem Eur J 13:3636–3642
29. Steiner A, Stalke D (1993) J Chem Soc, Chem Commun 444–446
30. Becker G, Beck HP (1977) Z Anorg Allg Chem 430:77
31. Becker G, Niemeyer M, Mundt O, Schwarz W, Westerhausen M, Ossberger W, Mayer P, Nöth H, Zhong Z, Dijkstrac P et al (2004) Z Anorg Allg Chem 630:2605
32. Steiner A, Stalke D (1993) J Chem Soc, Chem Commun 444
33. Steiner A, Stalke D (1995) Organometallics 14:2422
34. Henn J, Meindl K, Oechsner A, Schwab G, Koritsanszky T, Stalke D (2010) Angew Chem 122:2472
35. Johnson AW (1993) Ylides and imines of phosphorus. Wiley, New York
36. Gleria M, Jaeger Rd (2004) Phosphazenes – a worldwide insight. Nova Publishers, New York
37. Dehnicke K, Weller F (1997) Coord Chem Rev 158:103–169
38. Gilheany DG (1994) Chem Rev 94:1339–1374
39. Steiner A, Zacchini S, Richards PI (2002) Coord Chem Rev 227:193
40. Bickley JF, Copsey MC, Jeffery JC, Leedham AP, Russell CA, Stalke D, Steiner A, Stey T, Zacchini S (2004) J Chem Soc, Dalton Trans 989
41. Reed AE, Schleyer PvR (1990) J Am Chem Soc 112:1434–1445
42. Magnusson EJ (1993) J Am Chem Soc 115:1051–1061
43. Chesnut DB (2003) J Phys Chem A 107:4307–4313
44. Naito T, Nagase S, Yamataka H (1994) J Am Chem Soc 116:10080–10088
45. Restrepo-Cossio AA, Gonzalez CA, Mari F (1998) J Phys Chem A 102:6993–7000
46. Yamataka H, Nagese S (1998) J Am Chem Soc 120:7530–7536
47. Dobado JA, Martínez-Garzía H, Molina JM, Sundberg MR (1998) J Am Chem Soc 120:8461–8471
48. Lu CW, Liu CB, Sun CC (1999) J Phys Chem A 103:1078–1083
49. Lu CW, Sun CC, Zang QJ, Liu CB (1999) Chem Phys Lett 311:491–498
50. Koketsu J, Ninomiya Y, Suzuki Y, Koga N (1997) Inorg Chem 36:694–702
51. Kocher N, Leusser D, Murso A, Stalke D (2004) Chem Eur J 10:3622–3631
52. Wingerter S, Pfeiffer M, Baier F, Stey T, Stalke D (2000) Z Anorg Allg Chem 626:1121–1130
53. Wingerter S, Pfeiffer M, Murso A, Lustig C, Stey T, Chandrasekhar V, Stalke D (2001) J Am Chem Soc 123:1381–1388
54. Yufit DS, Howard JAK, Davidson MG (2000) J Chem Soc, Perkin Trans 2 2:249–253
55. Takeda T (2004) Modern carbonyl chemistry – methods and applications. Wiley-VCH, Weinheim
56. Aggarwal VK, Richardson J (2003) Chem Commun 21:2644–2651
57. Aggarwal VK, Winn CL (2004) Acc Chem Res 37:611–620
58. Brandt S, Helquist P (1979) J Am Chem Soc 101:6473–6475
59. Kremer KAM, Helquist P, Kerber RC (1981) J Am Chem Soc 103:1862–1864
60. O'Connor EJ, Helquist P (1982) J Am Chem Soc 104:1869–1874
61. Weber L (1983) Angew Chem 95:539–551

62. Aggarwal VK (1998) Synlett 4:329–336
63. Tewari RS, Awasthi AK, Awasthi A (1983) Synthesis 4:330–331
64. Franzen V, Driesen H-E (1963) Chem Ber 96:1881–1890
65. Stey T, Stalke D (2004) Lead structures in lithium organic chemistry. In: Rappoport Z, Marek I (eds) The Chemistry of Organolithium Compounds. New York: JohnWiley & Sons, pp. 47–120
66. Beswick MA, Wright DS (1998) Coord Chem Rev 176:373–406
67. Fleischer R, Stalke D (1998) Coord Chem Rev 176:431–450
68. Stalke D (2000) Proc Indian Acad Sci 112:155–170
69. Brask JK, Chivers T (2001) Angew Chem 113:4082–4098
70. Aspinall GM, Copsey MC, Leedham AP, Russell CR (2002) Coord Chem Rev 227:217–232
71. Walfort B, Leedham AP, Russell CR, Stalke D (2001) Inorg Chem 40:5668–5674
72. Walfort B, Stalke D (2001) Angew Chem 113:3965–3969
73. Walfort B, Bertermann R, Stalke D (2001) Chem Eur J 7:1424–1430
74. Deuerlein S (2007) Synthesis and electron density determination of novel polyimido sulfur ylides. Göttingen, Germany
75. Deuerlein S, Leusser D, Flierler U, Ott H, Stalke D (2008) Organometallics 27:2306–2315
76. Herberhold M, Köhler C, Wrackmeyer B (1992) Phosphorus, Sulfur, Silicon Relat Elem 68:219–222
77. Pohl S, Krebs B, Seyer U, Henkel G (1979) Chem Ber 112:1751–1755
78. Mayer I (1987) THEOCHEM 149:81–89
79. Bors DA, Streitwieser A (1986) J Am Chem Soc 108:1397–1404
80. Salzner U, Schleyer PvR (1993) J Am Chem Soc 115:10231–10236
81. Stefan T, Janoschek R (2000) J Mol Modeling 6:282–288
82. Cruickshank DWJ, Eisenstein M (1985) J Mol Struct 130:282–288
83. Cruickshank DWJ (1985) J Mol Struct 130:177–191
84. Rundle RE (1949) J Chem Phys 17:671–675
85. Rundle RE (1957) J Phys Chem 61:45–50
86. Steudel R (2008) Chemie der Nichtmetalle - Von Struktur und Bindung zur Anwendung, Berlin: de Gruyter
87. Fleischer R, Freitag S, Pauer F, Stalke D (1996) Angew Chem 108:208–211
88. Fleischer R, Rothenberger A, Stalke D (1997) Angew Chem 109:1141–1143
89. Fleischer R, Freitag S, Stalke D (1998) J Chem Soc, Dalton Trans 193–197
90. Fleischer R, Stalke D (1998) Chem Commun 343–345
91. Fleischer R, Stalke D (1998) J Organomet Chem 550:173–182
92. Fleischer R, Stalke D (1998) Organometallics 17:832–838
93. Ilge D, Wright DS, Stalke D (1998) Chem Eur J 4:2275–2279
94. Walfort B, Pandey SK, Stalke D (2001) Chem Commun 1640–1641
95. Fleischer R, Walfort B, Gbureck A, Scholz P, Kiefer W, Stalke D (1998) Chem Eur J 4:2266–2279
96. Glemser O, Pohl S, Tesky F-M, Mews R (1977) Angew Chem 89:829–830
97. Meij R, Oskam A, Stufkens DJ (1979) J Mol Struct 51:37–49
98. Herbrechtsmeier A, Schnepfel FM, Glemser O (1978) J Mol Struct 50:43–63
99. Markowskii LN, Tovestenko VI, Pashinnik VE, Mel'nichuk EA, Makarenko AG, Shermolovich YG (1991) Zh Org Khim 27:769–737
100. Leusser D, Henn J, Kocher N, Engels B, Stalke D (2004) J Am Chem Soc 126:1781–1793
101. Henn J, Ilge D, Leusser D, Stalke D, Engels B (2004) J Phys Chem A108:9442–9452
102. Braunschweig H, Kollann C, Rais D (2006) Angew Chem Int Ed 45:5254–5274
103. Braunschweig H, Kollann C, Seeler F (2008) Struct Bond 130:1–27
104. Braunschweig H, Kollann C, Englert U (1998) Angew Chem 110:3355
105. Braunschweig H, Colling M (2001) Coord Chem Rev 223:1–51
106. Braunschweig H, Kollann C, Rais D (2006) Angew Chem 118:5380–5400
107. Braunschweig H, Burschka C, Burzler M, Metz S, Radacki K (2006) Angew Chem 118:4458–4461

108. Braunschweig H (1998) Angew Chem 110:1882–1898
109. Braunschweig H, Colling M, Hu C (2003) Inorg Chem 42:941–943
110. Aldridge S, Coombs DL (2004) Coord Chem Rev 248:535–559
111. Braunschweig H (2004) Adv Organomet Chem 51:163–192
112. Ehlers AW, Baerends EJ, Bickelhaupt FM, Radius U (1998) Chem Eur J 4:210–221
113. Boehme C, Uddin J, Frenking G (2000) Coord Chem Rev 197:249–276
114. Blank B, Colling-Hendelkens M, Kollann C, Radacki K, Rais D, Uttinger K, Whittell GR, Braunschweig H (2007) Chem Eur J 13:4770–4781
115. Flierler U, Burzler M, Leusser D, Henn J, Ott H, Braunschweig H, Stalke D (2008) Angew Chem 120:4393–4397
116. Götz K, Kaupp M, Braunschweig H, Stalke D (2009) Chem Eur J 15:623–632
117. Farrugia LJ, Mallinson PR, Stewart B (2003) Acta Crystallogr B59:234–247
118. Leung PC, Coppens P (1983) Acta Crystallogr B39:535–542
119. Macchi P, Garlaschelli L, Martinengo S, Sironi A (1999) J Am Chem Soc 121:10428–10429
120. Farrugia LJ, Evans C, Tegel M (2006) J Phys Chem A 110:7952–7961
121. Haaland A, Shorokhov DJ, Tverdova NV (2003) Chem Eur J 10:4416–4421
122. Macchi P, Garlaschelli L, Sironi A (2002) J Am Chem Soc 124:14173–14184
123. Henn J, Leusser D, Stalke D (2007) J Comput Chem 28:2317–2324
124. Bochmann M (1992) Angew Chem 104:1206–1207
125. Piers WE, Chivers T (1997) Chem Soc Rev 26:345–354
126. Piers WE (1998) Chem Eur J 4:13–18
127. Stephan DW, Erker G (2010) Angew Chem 122:50–81
128. Kehr G, Fröhlich R, Wibbeling B, Erker G (2000) Chem Eur J 6:258–266
129. Flierler U, Leusser D, Ott H, Kehr G, Erker G, Grimme S, Stalke D (2009) Chem Eur J 15:4595–4601

Struct Bond (2012) 146: 101–126
DOI: 10.1007/430_2012_79
© Springer-Verlag Berlin Heidelberg 2012
Published online: 27 March 2012

Electronic Structure and Chemical Properties of Lithium Organics Seen Through the Glasses of Charge Density

Dirk Leusser

Abstract This chapter presents an overview on lithium organic compounds which have already been subject to topological analyses of the experimentally derived charge density distributions. The rich nature of the Li–C bond ranges from a polar 2c2e single contact in donor base stabilised lithiumorganic monomers to an anionic C_α capping a lithium triangle in a 4c2e Li_3–C_α bond in tetramers or hexamers. The presented examples will show how powerful and predictive the topological analysis on experimental data can be. The authors cited herein were able to classify various types of atomic interactions (multiple bonds, bond polarisation and dative bonds), give insight into charge separation within the molecules, identify hybridisation states, and explain or even predict the chemical reactivity.

Keywords lithium organics • experimental charge density • multipole refinement • topological analysis • electron density distribution

Contents

D. Leusser (✉)
Institut für Anorganische Chemie, Tammannstraße 4, 37077 Göttingen, Germany
e-mail: dleusser@gmx.de

1 Introduction

This chapter gives an overview of the topological analysis on *experimentally* derived charge density distributions of molecules from high-resolution X-ray data in the class of lithium organics. Generally, all efforts to determine molecular structures are undertaken to deduce chemical and/or physical properties. A straightforward structure–property relation is the ultimate goal and from a chemists point of view this relatively new experimental technique represents the cutting edge of this development. It allows us to deduce chemical properties from the detailed electron density distribution and thus the *prediction* of chemical reactivity instead of the post-synthetic explanations from the bond lengths and angles rooted solid-state structure determination.

Over the last decade the analysis of experimentally derived charge density distributions made a huge step forward – coming from small molecules, easy to crystallise and handle, mostly organic compounds, to more advanced materials which bear a vast variety of challenges. These range from the experimental handling to the derivation of the electron density distribution from the measured intensities. The focus of interest has moved from a rather purely *physical* point of view – the investigation of the principal relations between measured data and the derivation of the density distribution from multipole refinement [1] with subsequent analysis in terms of Bader's Quantum Theory of Atoms in Molecules (QTAIM) [2] – to a more *chemical* one, namely the answer of concrete chemical questions and the calculation of a manifold of physical properties from the experimental charge density distribution.

In the light of this historical background it is not surprising that it took more than three decades of progress in the field of experimentally determined charge density distributions until lithium organics came into play. These compounds are challenging in any aspect: very difficult to handle during the experiment due to an extreme reactivity and sensitivity to air coupled with a low scattering power. In addition, the electron density distribution of these compounds is very often difficult to model since the valence density of the lithium atoms is made up by a single, probably diffuse electron. Therefore, the density distribution and its physical and chemical properties have to be interpreted with extreme caution. Theoretical investigations on the other hand were much earlier available. Gatti et al. [3] presented their analyses of planar lithium clusters in 1987, even 2 years before Bader and MacDougall deduced the reactivity from a topological analysis of $LiCH_3$ [4]. Already in these early days of charge density investigations, lithium-containing compounds were in the focus of research. In the same year, Reed et al. presented an elaborate work on Li–X compounds to illustrate the use of the natural population analysis [5]. However, this chapter deals exclusively with the *experimentally* derived charge density distributions of lithium organics and their topological analysis and does not cover the theoretical investigations.

It was not until 2001 [6, 7] since the first Li-containing compound was investigated by means of a topological analysis based on experimental X-ray data, even though lithium compounds are by far the most elaborated subset in the field of s-block metal organics [8]. The interest in lithium organics is mainly based on three aspects. The first is their role in preparative chemistry (e.g. for the deprotonation of weakly acidic reagents, the transfer of organic groups, or anionic polymerisation), which always fuelled the hope for a structure–property relation and made them a playground for crystallographers all over the world's X-ray laboratories. This seems feasible since the technical requirements were established, e.g. cooling devices [9], area detectors and very intense X-ray sources even on in-house facilities. Second, the unique structural features, which originate from the tendency to build up deltahedral metal cores which aggregate from a Li_3 triangle μ_3-capped by a carbanionic C_α atom (Fig. 1). Thus, lithium organic compounds show a variety of Li–Li and Li–C contacts. Their properties or even existence is under debate for decades, and they are still a great scientific challenge today. The third aspect is strongly related to the structural features. Many of the lithium organics form exceptionally aesthetic compounds, which originates in the tendency of forming higher aggregates of the Li_3 subunits.

We can define two classes of investigations based on the experimentally derived charge density distribution on lithium-containing compounds in literature so far: Those with a lithium atom counterbalancing the charge of a formally charged ligand. For those where the nature of the donor–lithium bonds or the lithium–lithium contacts is in the focus of interest. On the other hand those molecules, where the charge distribution of the ligand itself and the connected reactivity and physical properties are under investigation.

Historically, it was the bonding of the Li–donor which was first examined by means of an experimental charge density distribution from high-resolution X-ray data.

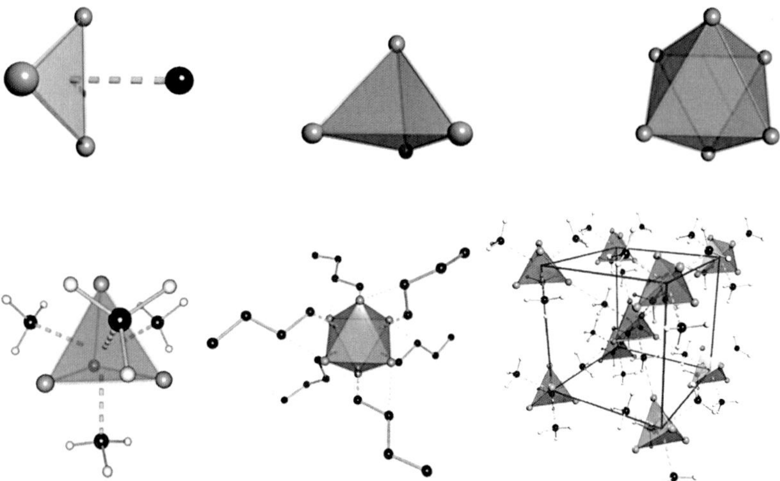

Fig. 1 Aggregation of the μ_3-capped Li_3-triangle to give deltahedral metal cores and the representative solid state structures of tetrameric $[t\text{-BuLi}]_4$, hexameric $[n\text{-BuLi}]_6$ and polymeric $[\text{MeLi}]_4$ (taken from [8])

A decade ago, Scherer et al. [6] were the first to give some insight into the charge density properties of a picolyl-lithium species in a combined theoretical and experimental study. However, their attention was mainly focused on an agostic interaction in question. Still, almost a decade later, there are just a few examples published, where the bonding of the lithium cation to a negatively charged counter ion was analysed in detail. Kocher et al. determined the nature of a Li–N bond which they classified to show predominantly ionic characteristics and the resulting impact on the density distribution in a formal hypervalent iminophosphorane $[(Et_2O)Li\{Ph_2P(CHPy)(NSiMe_3)\}]$ [10]. Their findings showed a surprising accumulation of charge at the formal ionic centres within the molecule. The same authors extended their investigations of Li–donor interactions to a series of octamethylcyclotetrasilazane (OMCTS) derivates and they were the first who were able to base predictions about the chemical reactivity on the density features observed in solid state. They could determine two types of chemically equivalent but electronically distinguishable protons from the charge density distribution of the starting material OMCTS. This allowed to calculate the experimentally derived electrostatic potential and from that, to deduce the chemical reactivity of OMCTS. This was proved by a transannular dimetallation to give a lithium complex containing the cyclic dianion [11].

Deuerlein et al. [12] and just shortly later Ott [13] went a step further and concentrated their attention to a potential Li–Li interaction in question. Both did not find any topological proof for the existence of a Li–Li linkage by means of an attractive Li–Li bond. In the same study, Deuerlein et al. were the first to identify the nature of the fundamental Li_3–C_α building block of lithium-organic deltahedral tetra- and hexamers. They found a carbon-centred lone-pair pointing towards the Li_3-triangle and in addition three paths of maximum density from the carbon atom to each of the lithium cations. This, together with the two hydrogen bonds and a sulphur bond of the SH_2C_α-Li_3 unit, led the authors to the classification of the carbon atom to be sixfold coordinated.

Ott et al. focused their investigations on the reactivities of the lithium-species in question. In a study on 2-picolyllithium [14] they could straightforwardly distinguish their compound to be an amide in contrast to the widely accepted picture of a carbanionic species, which explains the related reactivity, e.g. the electrophilic attack at the methyl position. In a study on a stereochemical benchmark system, (*R*,*S*)-**2**·quinuclidine, they were even able to predict the regioselective reaction pathway from its experimental density distribution [15].

The so far latest investigations on a hexameric trimethylsilylmethyllithium are a merge of the questions which arise in structure-based lithium-organic chemistry and show the strength of the topological analysis of experimentally determined charge density distributions [13]. Herein, the Li–Li distances and Li_3–C_α bonds as well as the agostic Li–H contacts in question and even structurally implied hyperconjugation could be studied in one single molecule. This may lead to answers for open questions in a decades-long debate, like *what keeps lithium organics together*? or *what is the nature of the Li–X interaction*? or *how can we predict the reactivity of lithium organics*? in one single experiment.

2 Methodical Overview

A short summary of the experimental and theoretical background, the applied tools, programs and formalism will be given. This overview will be restricted to those aspects, which are needed to follow the arguments of the published examples on lithium complexes cited here. Therefore, this should be taken by no means as a general synopsis for the method. For deeper insight the reader is referred to the comprehensive compendia by Coppens [1] and Bader [2].

2.1 Experimental Requirements

Application of a topological analysis based on experimental data requires extreme caution during data acquisition and crystal preparation. The data quality needed is by far superior to the standard maintained in the day-to-day lab work. Usually, lithium organics are extremely reactive making it necessary to prepare and select the crystals at cryogenic inert conditions with Schlenk techniques [9].

Since standard structure determination is based on 9 variables per anisotropic atom, this number exceeds up to 38 if the electron density is refined by a multipole model expanded to the fourth order. Simple mathematical considerations (based on the need of an over-determined system of equations for the least-squares routine) show, that at least four times as many integrated intensities as for a standard structure determination are needed for the refinement process. What makes it even more challenging is that this data has to be of superb quality. The only way to measure additional unique data is to extend the data collection to high scattering angles. As a general rule of thumb a minimum resolution of $(\sin \Theta/\lambda)_{\text{min}} = 1.0\ \text{Å}^{-1}$ is widely accepted. This limit is rarely reached for lithium organics due to their low scattering power, resulting from the absence of heavy elements. One could try to overcome this problem by increasing the scattering volume of the crystal. Unfortunately, this way of solving the problem is limited. Modern intense X-ray sources combine a very high photon flux which is focused on the sample, e.g. by use of a rotating anode with mirror optics instead of graphite monochromators [16, 17]. The high quality requirements make it necessary to keep a constant crystal volume exposed to the beam during the whole experiment. Beam profiles in the range of 100–300 μm can cause severe problems during data reduction and scaling if larger crystals were used [18].

Nowadays, it is common to collect data with an area detector. Then, it is essential to measure with high redundancy to correct for systematic errors like absorption. Together, all these requirements – high redundant high-resolution data of small crystals – lead to measurements lasting for weeks instead of days.

The temperature has to be kept as low as possible, to enhance the scattering power due to reduced atomic motion on the one hand and to prevent the compounds from thermal decomposition on the other hand, which is a widely observed behaviour especially for lithium organics. Lots of experiments of these reactive species have been reported to fail, due to thermal decomposition, reaction of the crystal with air, or icing problems. Icing is one of the main problems during long-term low-temperature experiments. Exposure times in the range of minutes at high angles with the crystal in unchanged position bathing in a nitrogen or helium cold gas stream, facilitates humidity to freeze out at the crystal. This is the potential end of any experiment. Some effort has been made to reduce the humidity at the crystal environment, mostly realised by drying the air with air-conditioner or freezer-like techniques [19]. Keeping a potentially unstable and reactive sample for weeks under perfect experimental conditions is one of the main challenges during the X-ray experiments with lithium organics. This might be one reason why relatively few examples are known so far in literature.

2.2 *Special Requirements for the Data Reduction*

Data processing has become a virtually automated process for standard structure determinations. Program suites like APEX2 [20] or WINGX [21] provide an easy to handle set of tools and programs which allow the user to integrate, scale and correct the data for absorption and other systematic errors, to determine the space group, and finally set up the files needed for structure solution and refinement in a guided step-by-step manner.

Data reduction of high-resolution data requires much care in every step of the process. Data integration as well as the scaling of the data is complicated by an inherent 2Θ-dependence of both, shape and intensity of the measured reflections. Different integration programs handle the angular-dependent broadening of the reflections with different algorithms, e.g. increase of the integrated volume with large scattering angles, profile-fitting versus simple-sum routines for weak reflections or the consideration of the $K_{\alpha 1}/K_{\alpha 2}$-split for high-resolution data. However, none of the commonly used integration programs can account perfectly for the systematic errors incorporated by a wide 2Θ-range. The result of such systematic bias in the data has enormous influence on the scaling factor and due to its high correlation with all refined parameters on the structural model itself.

The sources of errors listed above are of general type. However, for the class of lithium organics, especially the data reduction is a challenging issue, as due to the low scattering power all inherent problems seem to be amplified. All published structures suffer from a relatively low resolution during data acquisition. Even if the resolution limit of $(\sin\Theta/\lambda)_{\min} = 1.0\ \text{Å}^{-1}$ is reached, the data quality decreases rapidly with increasing scattering angle. The proportion of weak data rises dramatically, making it necessary to model intensity profiles even if the experimental strategy is based on a fine-slicing technique which refers to simple-sum integration instead of profile-fitting routines. Especially if programs are used which use both, simple-sum and profile-fitting routines (e.g. SAINT [22]) the numerous proportions of weak data seem to introduce systematic bias. For these data sets in addition to the 2Θ-dependence originating from reflection broadening ($K_{\alpha 1}/K_{\alpha 2}$-split, incident angle effect), one has to take into account scaling factor dependence from the use of different integration algorithms combined for a single data set and therefore one single scaling factor applied to all data during the subsequent refinement. Integration programs which use exclusively one algorithm, like XDS [23] (profile fitting) or EVAL [24] (ab initio calculation of 3d profiles from physical crystal and instrument parameters) seem to suffer less from this bias. Comparison of published GoF-values at otherwise similar quality criteria demonstrates this impression.

Virtually originating in the same principal shortcomings, problematic estimation of standard deviations, especially for weakly scattering samples is observed. On the one hand, high redundancy is required to determine reliable intensities, on the other hand this leads to underestimated standard deviations after the merging routine. A speciality of many lithium organics, e.g. those of synthetical interest, is their chirality, leading to chiral space groups making the whole procedure even more challenging, since it complicates data collection with high redundancies and completeness and halves the number of symmetry equivalents needed for empirical approaches to correct for absorption. Two programs are used in the presented examples, SADABS [25] and SORTAV [21]. Both are combining absorption correction (numerical based on face indexing or multi-scan based on the Blessing algorithm [26]), scaling and merging. Data integrated with SAINT is mostly corrected by SADABS, while users working with XDS or EVAL tend to prefer SORTAV. On the first glance, data corrected with SADABS seems to suffer more from an underestimation of the standard deviation, which can be deduced from systematically higher GoF values. However, the author addresses this effect to the integration procedure instead the subsequent absorption correction and merging. SABABS users may "transfer" the bias introduced by the integration software (namely SAINT) to a higher extend into the refinement process, since during correction and merging with SADABS no arbitrary corrections are made in addition to those exactly deducible from statistics. Independent from the data "quality" in the sense of exact measurement of the intensities, standard deviations estimated by other programs than SAINT seem to be less optimistic and therefore "more reliable" even though larger in value.

Whatever program suite or mathematical approach is used, one has to keep in mind that especially for weak scattering samples bias is introduced, which manifests first of all in the scaling factor, the refinement parameter with most prominent influence. Therefore, special care has to be taken not just during data reduction, but especially during refinement and the analysis of the refinement statistics. Here, one might get more insight into the exact coherence to the 2Θ-dependence and be able to correct the data and the estimated standard deviations in an iterative procedure [27].

2.3 *Structure Refinement: Modelling Li^+–X^-*

All lithium organic compounds presented here were refined using the Hansen and Coppens multipole model [28] implemented in the program package XD or XD2006 [29]. Therefore, the details on the special requirements during structure refinement are restricted to the experiences of the author made with the program XD or those mentioned in the cited publications as long as the multipole modelling is described.

Concerning the principal procedure of aspherical modelling, lithium organics are treated the same way as other classes of compounds. In a first step a starting model (*i*ndependent *a*tom *m*odel, IAM) has to be refined in such a way, that the best starting atomic positions and thermal motion parameters for all atoms are determined, eventually at the cost of higher R-values. Usually, this is achieved by a high-order refinement for the non-hydrogen atoms and modelling of the hydrogen atoms by the use of low-order data combined with a riding model for the isotropic thermal motion parameters including distance constraints for the X–H bonds by fixing them to neutron distances [30]. This strategy leads to the best estimate of coordinates and thermal motion parameters from spherical structure factors.

It gets much less straightforward as soon as an aspherical modelling is introduced. Two main problems occur: first, if the space group is non-centrosymmetric, the refinement has to be stabilised by as many (chemically and physically substantiated) chemical and symmetry constraints as possible, the complexity of the model should eventually be restricted (maximum order of multipoles, kappa-restrictions) and a constraint should be introduced fixing the origin when a polar axis is present. This is automatically applied by a "floating origin" restraint (centre of gravity is fixed in the polar axis direction) during refinement with, e.g., SHELXL [31], but not in XD. The only way to simulate this is by fixing one atom's coordinate on the polar axis. In that case the possible problem of the "floating origin" is prevented, but on the cost of higher correlations. In addition to these correlations, fixing of a coordinate is a rather arbitrary proceeding, especially when no heavy atoms are present, which is the case in all published examples.

The second limiting factor is again strongly related to the low scattering power of lithium organics. The low number of strong high-resolution data also makes it necessary to refine constrained models of reduced flexibility. To keep the system of linear equations during the least-squares refinement process over-determined to a sufficient degree, one has to introduce constraints for the multipole populations based on approximate chemical equivalence as well as non-crystallographic symmetry restrictions (only those multipoles are refined, which fulfil the local symmetry) and one might be forced to reduce the maximum order of the multipolar expansion of the aspherical density.

Lithium organics often consist to a high percentage of hydrogen atoms. The structures of $[t\text{-BuLi}]_4$ and $[n\text{-BuLi}]_6$ (Fig. 1) might serve as a reference in this aspect. In tetrameric $[t\text{-BuLi}]_4$ already one fourth of the scattering electrons are hydrogen-centred and this ratio even increases to 50% in hexameric $[n\text{-BuLi}]_6$. The limited success in the description of the aspherical hydrogen densities is well understood and one can try to minimise the errors introduced herby by using very restricted models for the hydrogen atoms, riding models for the thermal motion, and distance constraints. However, it is the principle lack of information about the diffuse electron density distribution of the hydrogen atoms deducible from the measured intensities which limits the quality of the refined model. This limit is even more distinct the higher the percentage of hydrogen atoms in the structure is and at ratios of 50% one has to ask oneself whether this principle limit might affect the quality of the description of the non-hydrogen atoms, too. To prevent the reader to yield to despair, a very promising approach should be mentioned here. Overgaard et al. with the program SHADE [32] and Dittrich et al. [33] with the INVARIOM approach presented very promising tools to describe hydrogen atoms with an anisotropic thermal motion model, which for the first time allows to model reliable aspherical densities for hydrogen atoms without information from neutron data. Then, we can hope that the modelled aspherical density can be deconvoluted to a sufficient degree from its smearing due to thermal motion.

An effect related to that originates in the very special structural properties of lithium organics. Many of them consist of lithium triangles capped by C_nH_m functional groups. Those are often chain-like arranged or perimeters with high rotational freedom. Those groups are "fixed" to the lithium triangles via one terminal (carbon) atom. This allows a high flexibility to give rotational or displacement disorders. This structural characteristic might be the origin of the low scattering power, which was identified to be the root for many problems, while working with lithium organics.

Weak scattering and its impact on the scaling factor were already described above. Strongly related to the scaling factor is the used weighting scheme during refinement. IAM-based programs like SHELXL comprise a semi-automated (by suggestion) adjustment of the weighting scheme. Although the weighting scheme is the same in XD, the automation is not implemented. However, the weighting scheme suggested by SHELXL should not straightforwardly be used during multipole refinement, since it is calculated to give a flat analysis of variance and an aspherical refinement describing those features which create errors in spherical models (interatomic residual densities), will for sure need a different weighting. Therefore, we have to be very careful by using this corrective tool to account for 2Θ-dependent systematic errors. A solution to the problem might be a careful analysis of the statistics subsequent to a multipole refinement (e.g. XDWTAN implemented in XD) as it was presented by Pinkerton et al. [27] However, one has to keep in mind that this is a post-diagnostic procedure and a proper weighting scheme deduced after refinement has to be tested and probably adjusted in an iterative way – yet another time-consuming step.

Typical even though not unique for lithium organics, we have to consider one structural inherent question for all zwitterionic species. The usage of ionic scattering factors (in this case Li^+) is provided by XD in a straightforward manner. Nevertheless, problems occur if one does so, since as long as the charge is balanced by the organic framework, no well-defined anion can be addressed and the electroneutrality is violated. Different treatments were described in the literature. While Scherer et al. [6] seemed not to pay attention to this "artificial" charge and prevented from refining a pseudo-minimum by introducing the "keep charge" constraint, Deuerlein et al. [12] and Kocher et al. [11] avoided any charge preassumption by using the neutral scattering factor and refining the Li-monopole, while Ott et al. [14, 15] describe a procedure where they distribute the negative charge over the possible centres of negative charge and leave it to the refinement (monopole population with an electroneutrality constraint) where the cation is located. For the author's taste and based on his own experience it seems to be preferable to refine the monopole populations of neutral lithium atoms if possible. This introduces least bias and does not imply any pre-conclusion. Unfortunately, this is by far not always possible. Monopole populations of lithium atoms tend to adopt non-physical (negative) values and/or the associated expansion/contraction parameters refine to extremely large values, indicating a contraction (comparable to artificially high or low thermal motion parameters in the IAM) which overcompensates the loss of electrons due to ionisation. In this case restrictions to the expansion/contraction parameters have to be applied and if the model tends to lead to zero population, it is justified to use the Li^+ scattering factor, but then electroneutrality should be introduced "by hand" as suggested by Ott et al.

It is left to the imagination of the reader, if it is the sum of all the problems and pitfalls listed above which might be the reason for the relatively small number of lithium organic investigated so far or if it is the lack of open questions in this class of compounds accessible by the analysis of the experimental charge density. Hopefully Chap. 3, summing up which chemical and physical properties we can deduce so far from the experimental charge density, will give the answer.

2.4 Topological Analysis

A very brief summary of some topological criteria is given here.[1] Only those descriptors of R.W.F. Bader's QTAIM will be mentioned which were used in the publications cited here. For a more elaborated description the reader is referred to other chapters within this book, where the theoretical and methodical background is described in great detail.

Following QTAIM, the chemical structure of a molecule can be extracted from the analysis of the topology of $\rho(\mathbf{r})$, namely its *critical points* and the *curvatures* there. Local extrema, the so-called critical points occur where

$$\nabla\rho(\mathbf{r}) = \mathbf{i}\,\partial\rho/\partial x + \mathbf{j}\,\partial\rho/\partial y + \mathbf{k}\,\partial\rho/\partial z$$

vanishes, with the unit vectors **i**, **j** and **k**. Rank (w) and signature (σ), calculated from the eigenvalues of the diagonalised Hessian matrix, $\mathbf{H}(\mathbf{r}) = \partial^2\rho/\partial x_i\partial x_j$, classify the type of critical points (w, σ), e.g. atomic position (local maximum (3,−3) critical point) or *b*ond *c*ritical *p*oint (BCP, (3,−1) saddle point).

[1] This short overview is very much inspired by the chapter "Synopsis for non-crystallographers" of the XD2006 user's manual.

Analysis of the gradient of $\rho(\mathbf{r})$ leads to a characterisation of bonds by the inspection of the b*ond* p*ath* (BP), the line of maximum density between two nuclei, in terms of its length and bending, and the topological criteria at the b*ond critical* p*oint* (BCP), like its position on the path relative to the two bonded atoms, the density, $\rho(\mathbf{r}_{BCP})$, the sum of the three eigenvalues of the Hessian matrix, $\nabla^2\rho(\mathbf{r}_{BCP})$, and the ellipticity, $\varepsilon(\mathbf{r}_{BCP})$, the ratio of the two curvatures of the density perpendicular to the bond.

General rules facilitate the classification of bonds via the topological criteria at the BCP. Strength and multiple bond character rises with $\rho(\mathbf{r}_{BCP})$, negative $\nabla^2\rho(\mathbf{r}_{BCP})$ is typical for shared, positive values for closed shell interactions, non-zero ellipticity can be caused by π-contributions or coupling of lone-pair density into the bond. Polarisation of bond density manifests in a shift of the BCP towards the more electropositive bonding partner.

The *gradient vector field* of the charge density, the representation of trajectories of $\nabla\rho(\mathbf{r})$ and its critical points, is a topological visualisation of the molecule, in terms of atomic basins defined by the zero-flux surface, the bond paths, atoms (attractors of the trajectories) and BCPs. It can serve as reference if possible interaction pathways, size and shape of atoms are to be analysed.

Atomic interaction and the appendant rearrangement of the outer valence shell in *v*alence *s*hell *c*harge *c*oncentrations (VSCC) and *d*epletions (VSCD) is directly represented by the trace of the Hessian matrix, the *Laplacian* $\nabla^2\rho(\mathbf{r})$. Negative values refer to local concentrations, positive values to local depletions. The Laplacian distribution recovers the electronic shell model of an atom and serves as a faithful tool to identify for example non-bonding VSCCs (lone-pairs), bond-induced or ligand-induced CCs, shared versus closed-shell interactions, a Lewis base from a Lewis acid, and possible reactive sites for nucleophilic attack.

2.5 Chemical and Physical Properties from the Charge Density

Once the charge density distribution $\rho(\mathbf{r})$ is determined, a variety of density-related properties can be calculated. Examples published in the context of lithium organics so far are:

The e*lectrostatic* p*otential* (ESP), $\Phi(\mathbf{r})$

$$\Phi(\mathbf{r}) = \sum_j \frac{Z_j}{|\mathbf{r} - \mathbf{R}_j|} - \int \frac{\rho(\mathbf{r})}{|\mathbf{r} - \mathbf{r}'|} \mathrm{d}\mathbf{r}',$$

where $\mathbf{R}_j$ and Z_j are the position and charge of the j-th nucleus, respectively. The three-dimensional representation can point to possible sites for electrophilic attacks by inspection of regions with negative ESP.

The *electronic energy density* $E(\mathbf{r})$ of the electron density distribution is defined as $E(\mathbf{r}) = G(\mathbf{r}) + V(\mathbf{r})$, where $G(\mathbf{r})$ is a local one-electron kinetic energy density and $V(\mathbf{r})$ is the potential energy. With

$$L(\mathbf{r}) \equiv -(\hbar^2/4m)\nabla^2\rho\,(\mathbf{r})$$

we get

$$-\,L(\mathbf{r}) = (\hbar^2/4m)\nabla^2\rho\,(\mathbf{r}) = 2G\,(\mathbf{r}) + V(\mathbf{r}),$$

a unique relation of a property of the electronic charge density to the local components of the energy density. Several approximations to determine $E(\mathrm{r})$ from the experimental charge density distribution were suggested, the one given by Kirzhnits [34] is widely used and is a straight criterion for the recognition of the atomic interaction type: $E(r) < 0$ at the BCP is observed in shared-type (covalent) atomic bonding, while $E(r) \geq 0$ is observed in purely closed-shell (ionic) interactions [35].

In the quantum chemical framework of QTAIM, discrete atomic volumes, the atomic basins, are defined by the zero-flux surface, where $\nabla\rho(\mathbf{r})\cdot\mathbf{n}(\mathbf{r}) = 0$, with $\mathbf{n}(\mathbf{r})$ is the normal vector to the surface. This allows the determination of quantum chemically based *atomic charges* by integration of the density over the atomic basins. These charges and the associated atomic volumes are determined by both, ionisation due to charge transfer from one atom to the other and the loss of charge density due to bond polarisation and the related shift of the basin boundary towards the more electropositive atom. The determination of charges was extended to what the authors called "group charges" [15], summing up the integrated charges of atomic contributions to molecular fragments. Those can serve as a sensitive measure for polarisation effects and can be extremely helpful to identify the anionic fragments in the investigated lithium organics.

In their investigation on a lithium sulphur ylide Deuerlein et al. [12] suggested the use of the VSCCs to calculate an alternative *geometry* of the molecular structure, instead of using the direct connection lines between the atomic cores. When they compared the bond angles to those made up by the local maxima in the negative Laplacian distribution, they found those made up by the VSCCs to suite almost perfectly to the ones deduced by hybridisation concepts, while the standard bond angles did not allow any insight into the electronic state of the involved atoms.

3 Chemical, Physical and Structural Interpretation of the Charge Density Distribution in Lithium Organics

Due to the relatively few examples published so far, all of them can be presented here. Figure 2 gives an overview over the molecular structures of lithium organics of which experimental charge density data are available.

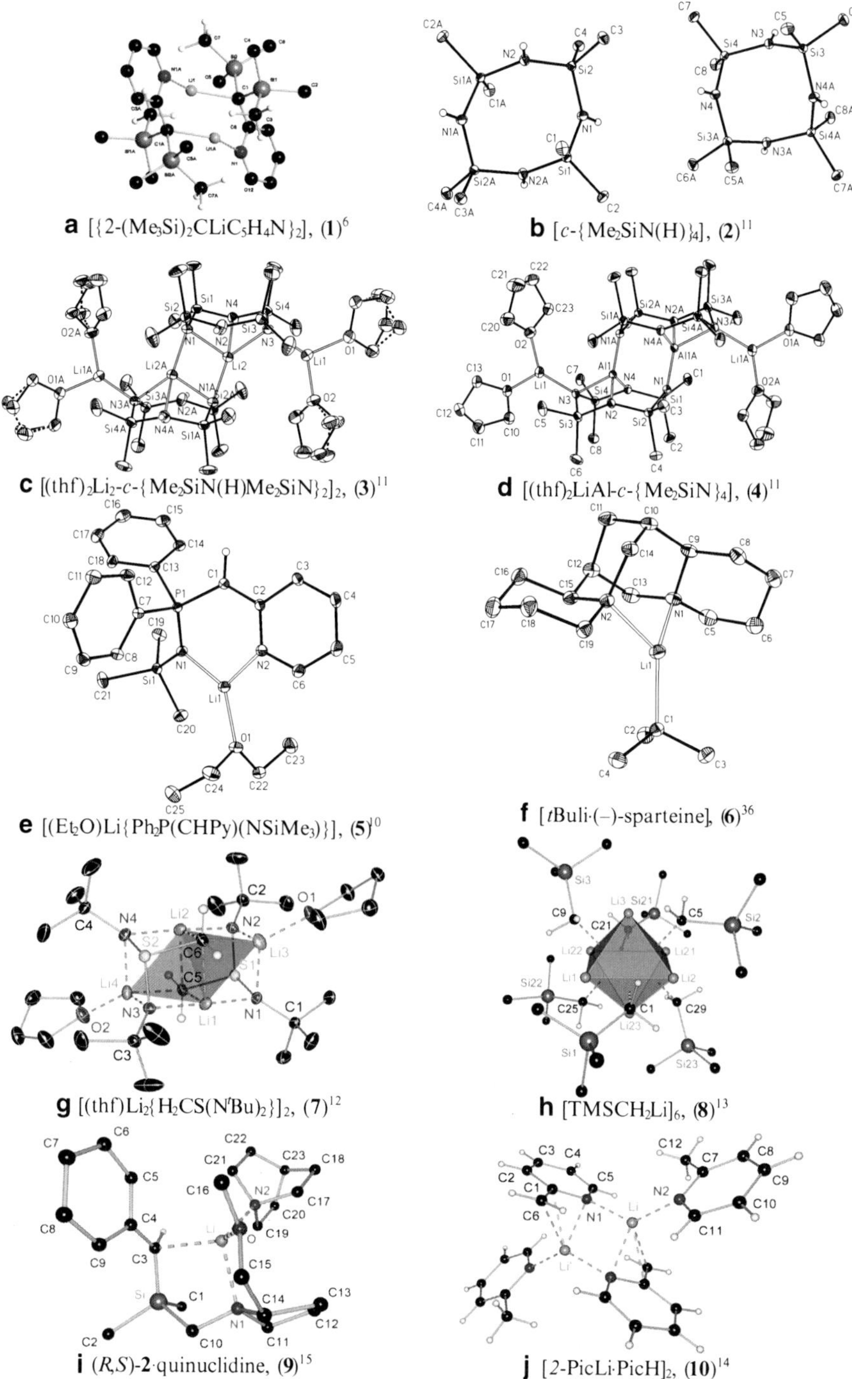

a $[\{2\text{-}(Me_3Si)_2CLiC_5H_4N\}_2]$, (**1**)[6]

b $[c\text{-}\{Me_2SiN(H)\}_4]$, (**2**)[11]

c $[(thf)_2Li_2\text{-}c\text{-}\{Me_2SiN(H)Me_2SiN\}_2]_2$, (**3**)[11]

d $[(thf)_2LiAl\text{-}c\text{-}\{Me_2SiN\}_4]$, (**4**)[11]

e $[(Et_2O)Li\{Ph_2P(CHPy)(NSiMe_3)\}]$, (**5**)[10]

f [*t*BuLi·(–)-sparteine], (**6**)[36]

g $[(thf)Li_2\{H_2CS(N^tBu)_2\}]_2$, (**7**)[12]

h $[TMSCH_2Li]_6$, (**8**)[13]

i (*R*,*S*)-**2**·quinuclidine, (**9**)[15]

j $[2\text{-}PicLi{\cdot}PicH]_2$, (**10**)[14]

Fig. 2 Molecular structures of lithium organics for which experimental charge density data are available. The graph of (**a**) has been created by a res-file extracted from the CCDC, (**f**) and (**h**) with friendly permission from the PhD thesis of the respective authors, the other structures were taken

Not more than nine examples are known to the best of the author's knowledge, of which two were solely published in PhD thesis so far (probably due to the limited resolution or redundancy of the experimental data). Nevertheless, a variety of structural motifs, which range from monomers (**5**, **6** and **9**) and dimeric forms (**1**, **3**, **4** and **10**) to the typical deltahedral lithium-cores, namely edge-linked triangles in **7** and an octahedron in **8**, were investigated. The coordination of the lithium cations is mainly accomplished by nitrogen atoms of the anionic organic counterions, occasionally complemented by donating solvent molecules (thf) or by formal carbanions. The compounds investigated so far can be divided into three groups:

1. Li^+–R^- species, where the cation is coordinated by nitrogen atoms of the formally negatively charged counterion (**1**) and the coordination is completed by solvent molecules like thf or diethylether (**3** and **4**). The metal complexes of the di- and tetraanion of cyclotetrasilazane belong to this group, as well as the alkyllithium complex (**1**) and the lithiated iminophosphorane (**5**). Compounds of this group will be referred to as Li^+R^--type below. *t*Buli·(–)-sparteine (**6**) is the link between this group and those species, which contain a formal carbanion.
2. In **2** and **3** Li^+ is bonded to the lone-pairs of neighboured nitrogen atoms by so-called dative bonds and to a formal carbanion of differing type. The formal C^-–Li^+ bonds are made up by R_2HC^- in the monomer (**9**) and by RH_2C^- (**10**) in the dimer. These compounds will be summarised as C^--type structures.
3. Deltahedral Li-complexes, named *δLi-type* below, forming a distorted Li-octahedron in **8** with six of the eight lithium-triangles μ_3-capped by RH_2C^- and two edge-linked lithium triangles in **7**, which are μ_3-capped by CH_2.

According to this ordering scheme the structural and electronic properties will be presented here to allow a direct comparison.

3.1 The Charge Density Distribution in Structures of the Li^+–R^--Type

The alkyllithium complex $[\{2\text{-}(Me_3Si)_2CLiC_5H_4N\}_2]$, (**1**) [6], the cyclotetrasilazanes (**3** and **4**) [11], and the lithiated iminophosphorane (**5**) [10] belong to this class. The distribution of charge over an anionic backbone is immanent to all the published structures of this group. Some effort has been made to identify the centres of charge localisation and the impact primarily on the bonding properties. Scherer et al. focused

Fig. 2 (continued) from the cited publications. **2** is not a lithium organic but cited here, since its reactivity to give **3** was proved by a charge density investigation. (**a**) $[\{2\text{-}(Me_3Si)_2CLiC_5H_4N\}_2]$, (**1**) [6]; (**b**) [*c*-$\{Me_2SiN(H)\}_4$], (**2**) [11]; (**c**) $[(thf)_2Li_2$-*c*-$\{Me_2SiN(H)Me_2SiN\}_2]_2$, (**3**) [11]; (**d**) $[(thf)_2LiAl$-*c*-$\{Me_2SiN\}_4]$, (**4**) [11]; (**e**) $[(Et_2O)Li\{Ph_2P(CHPy)(NSiMe_3)\}]$, (**5**) [10]; (**f**) [*t*Buli·(–)-sparteine], (**6**) [36]; (**g**) $[(thf)Li_2\{H_2CS(N^tBu)_2\}]_2$, (**7**) [12]; (**h**) $[TMSCH_2Li]_6$, (**8**) [13]; (**i**) (*R*,*S*)-**2**-quinuclidine, (**9**) [15]; (**j**) [2-PicLi·PicH]$_2$, (**10**) [14]

on the potential agostic interactions of the C_α–Si_β–C_γ–H_γ-backbone in **1** by inspection of the geometries, e.g. the bond lengths in the backbone and a detailed bond path analyses in their first short communication and gave a more detailed picture later in a comprehensive combined experimental and theoretical study [7]. The atomic model during the refinement of the lithium atom was quite restrictive. The lithium atom was treated as a cation with no electrons in the valence shell. Nevertheless, the bonding properties reflected in the bond lengths and the respective critical point properties allowed them to identify the interaction of second-order type due to delocalisation of charge over the backbone instead of a C–$H_\gamma\cdots Li^+$ agostic interaction. Most of their findings were based on indirect reasoning not by direct Li–H or Li–N bond analyses. No direct conclusion could be drawn from the atom distances alone. Li–N was in the range for standard dative bonds (1.9508 Å), Li$\cdots H_\gamma$ was at the short end of the range (2.043 Å), and C_γ–H_γ not elongated (1.087 Å). However, from the geometry, and first of all the extremely acute Li–C_α–Si angle (85.3°), an electronic reason had to be assumed. The authors identified it to originate in the density features of the C_α–Si_β–C_γ–H_γ-backbone. Taking into account that **1** was published already a decade ago, the fine details and the carefully drawn conclusions, which by the way hold until today without restriction, are superb. The density at the BCP in Li^+–C_α was reduced, while in the backbone it was found to be increased in C_α–Si_β compared to C_γ–Si_β, and unspecific in the C_γ–H_γ bond. From inspection of the BCPs, the authors went a step further and investigated the ellipticities along the whole bond paths (Fig. 3a). These profiles became a standard tool for the analyses of bonding types. However, the course of the ellipticities in **1** enlightened the charge delocalisation into the C_α–Si_β bond. This finding was further substantiated by the two-dimensional distribution of the Laplacian $L(\mathbf{r})$ (Fig. 3b). The VSCCs at C_α were found to merge which was taken as indication for delocalisation of charge from the Li–C bond into the backbone. Therefore, the observed geometrical characteristics were related to a negative hyperconjugation effect instead of a C–H bond activation and agostic interaction.

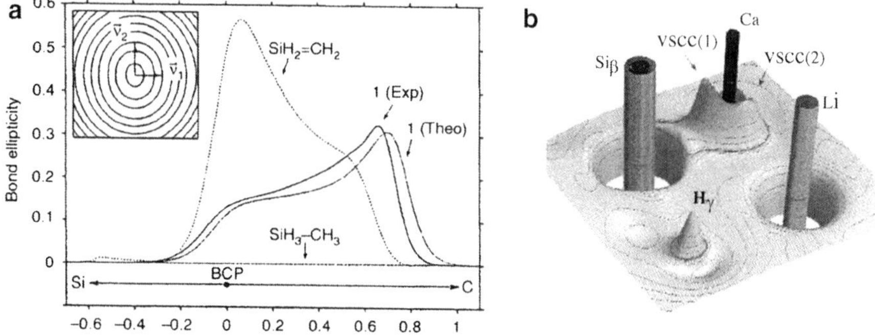

Fig. 3 Bond ellipticity along the C_α–Si_β bond (**a**) and relief map in the Li–C_α–Si_β plane (**b**) (taken from [6])

Kocher et al. used the same tools, e.g. bond path analyses by inspection of topological criteria along the whole bond, but they drew their attention on the nature of the Li–E (E = N, O) and Si–N interactions with and in the ligand in their study on octamethylcyclotetrasilazane and two lithium organic complexes containing the di- and tetraanion [11]. In addition to the bond analysis, they investigated the three-dimensional Laplacian distribution, the integrated charges, and calculated the electrostatic potential from the experimental charge density distribution. Their very elaborate comparison of three compounds and all corresponding bonds led to a conclusive picture of the inspected interactions. Small changes of the charge density distribution originating from the different deprotonation states were discussed. The major benefit of this study however was rooted in the chance to follow the electronic changes of the parent molecule during synthetic manipulation. The charge density distribution of OMCTS, **2**, represents in this context the undisturbed standard, while the lithiated dianion, **3**, and the mixed-metallated (Li, Al) tetraanion, **4**, allow to follow the impact of two- and fourfold deprotonation and subsequent metal coordination.

It was shown that the electronic properties were not reflected by bond lengths and angles. Insight into the nature of the Li–E bonds was given by a comparison of the Laplacian and ellipticities along the respective bond paths. All Li–E bonds were found to be extremely polar, classified as mainly ionic and very fine differences could be visualised especially by inspection of $L(r)$ along the whole bond. A very interesting speciality in the course of the Laplacian along the BP was not discussed by them, but might be seen in new light, now 6 years later. When they tried to identify the impact on deprotonation and metallation, they focused on bonds of the same type. They found the principal shape of the different Li–N, Li–Al and Li–O bonds being more or less the same, with fine changes mainly in the position and markedness of the minimum at the electronegative atom and the "flatness" in the interatomic region. What they did not discuss but might be of interest now, that we have more examples to compare, is that the course of the Li–O bonds was unique (Fig. 4). In contrast to the other two types it showed an even more distinct depletion

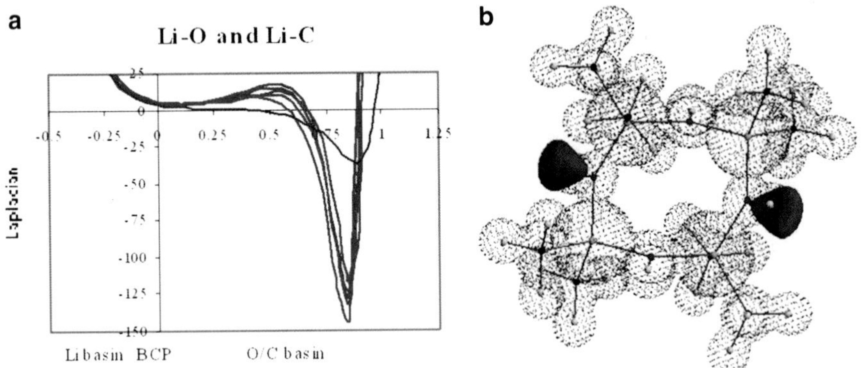

Fig. 4 $L(\mathbf{r})$ along the Li–E bonds (E = C, O) (**a**) and ESP in OMCTS (**b**) (taken from [36])

between the atoms already in the basin of the oxygen atom, but also a second minimum – even though on depletion level. This feature in the course of $L(\mathbf{r})$ seems to be a constant, which can be seen from the identical distribution found later in **5** [10]. Taking into account the attempts of the authors and other groups [37, 38] to discriminate dative bonds from ionic bond or covalent bonds, this shape might be an indicator for further investigations.

The deprotonation and subsequent metallation of OMCTS did not lead to increased ionicity of all Si–N bonds as could be expected from general rules. The authors found the lithium coordinated nitrogen atoms being higher charged, the respective Si–N bonds shortened, and the nitrogen lone-pairs less distorted compared to OMCTS. The lone-pairs at the deprotonated nitrogen atoms were found to be distorted, the VSCCs enlarged, and the density in the bond increased even though the bond path analysis revealed high polarity and ionic character which was attributed to a redistribution of lone-pair density into the bond.

The most striking finding in this publication was the correlation which could be drawn between the electrostatic potential calculated from the experimental density distribution and the reactivity of OMCTS during deprotonation and metallation reactions. Even though not forced by refinement constraints, the transannular hydrogen atoms in the eight-membered heterocycle in boat conformation adopted totally different values of the ESP. Just two of the four nitrogen bonded (transannular) hydrogen atoms showed regions of negative ESP and those are the most reactive sites where the deprotonation occurs first to give **3**. The two remaining hydrogen atoms are abstracted in a further synthetic step by $LiAlH_4$ leading to **4**. This is a very convincing example to show the *predictive* power of the experimental charge density analyses in metal organic chemistry.

Less eye-striking but with the same target to use the charge density distribution to predict and explain chemical reactivity, the same authors analysed a lithiated iminophosphorane, **5** [10]. The focus of interest in $[(Et_2O)Li\{Ph_2P(CHPy)(NSiMe_3)\}]$ was set on the Li–E (E = N, O) and the so-called "hypervalent" formal P=N and P=C bonds. The findings were very straightforward. Hypervalency was ruled out, the formal P=E (E = N, C) were described as P^+–E^- – single bonds, reinforced by electrostatic (ionic) contributions – and the Li–E (E = N, O) bonds were described as donor–acceptor bonds, with the lithium atom acting as acceptor of a "bifurcate" donation of two lone-pairs. Distinct negative charges, higher than expected from pure polarisation effects (up to −1.98 e), were found at the nitrogen atoms and at the phosphorous bonded carbon atom, counterbalanced by a huge positive charge of +2.20 e at the phosphorous atom and the positive charge at the lithium atom (explicit value not given in the paper). This, together with the results of the bond path analyses, which gave quite low $\rho(\mathbf{r}_{BCP})$ values for all P–E bonds and an even positive $\nabla^2\rho(\mathbf{r}_{BCP})$ for the formal P=C bond, made the authors reasoning the untypical reactivity – the P–E bonds are astonishingly easy to cleave for a double bond by metal organics in polar sovents – might have its origin in the widely accepted but false classification of these "hypervalent" bonds. The structure was therefore described as a zwitterionic phosphonium amide.

As a secondary finding – the main focus was not the lithium atom – a feature was mentioned, which seems to be another constant for lithium organics. The lithium atom tends to act as acceptor of bifurcated lone-pair donation. The lone-pair associated charge concentrations at the nitrogen atoms as well as at the oxygen atom in **5** are both oriented towards the lithium cation, forming acute angles but can still be resolved as individual concentrations leading to critical points in the Laplacian distribution (local maxima in $-L(\mathbf{r})$).

To understand the chemical reactivity seemed to be the driving force for the investigation of the [*t*Buli·(–)-sparteine], **6**, too [36]. **6** is one of the few examples of a monomeric lithium organic reagent, which is structurally characterised by means of X-ray analysis. Those compounds are believed to be the rate-determining species in lithium organic chemistry. The question here is the potential site for additional coordination of a nucleophilic reagent. Comparable to the findings in OMCTS, the electrostatic potential calculated from the experimental charge density distribution gives insight into possible reaction paths. The sites of negative electrostatic potential are exclusively found around the bulky ligand. Therefore, the lack of an additional ligand in the monomer is uniquely referred to the steric requirements. The three-dimensional representation of the ESP around the lithium atom allowed to preclude that the accessible area at Li^+ is sterically and electronically suited for additional coordination.

3.2 *The Charge Density Distribution in Structures of the C^--Type*

[*t*Buli·(–)-sparteine], **6**, is the link of the species described above to the C^--type molecules. CH_3 is bonded to the cationic lithium atom. The bond path analysis led to values for all Li–E (E = N, C) bonds which comply with the picture of dative lone-pair driven bonds dominated by ionic contributions. All three electronegative donating atoms exhibit qualitatively identical features. Distinct charge concentrations are found in the direction of Li^+, distorted at the carbanion, symmetrically distributed at the nitrogen atoms (Fig. 5). The charges are distinctly negative, but smaller than expected for the carbanion (−0.52 e), taking the three polarised hydrogen atoms into account. The bond path analysis underlines the idea of dominating ionic contribution to the bonding, with low densities at the BCPs accompanied by positive $\nabla^2\rho(\mathbf{r}_{BCP})$ and small but positive energy density.

More than 6 years lie between these investigations and the next publication of reactivity studies on carbanionic Li^+ species. Ott et al. presented two examples, one monomer [15] and a dimer [14] with a careful analysis each, leading to stringent explanation of the reactivity of these compounds. Their findings were commented in the same issue by Macchi in a "highlight" article, where some more theoretical details were also given [39]. The analysis of the dimeric 2-picolylithium targeted to answer two questions: (1) Should the compound be characterised as a carbanion or an amide? (2) What is the driving force for the reactivity?

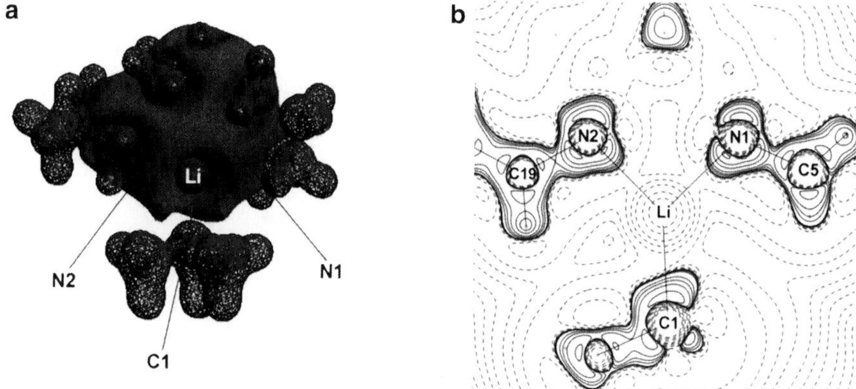

Fig. 5 ESP (**a**) and $L(\mathbf{r})$ (**b**) in **6** (taken from [36])

Both questions could be answered. The authors classified **10** as enamide. The ellipticities were analysed along the BPs and distinct π-contributions found in the anionic ring. The increased bond density was also reflected in the $\rho(\mathbf{r}_{\mathrm{BCP}})$ and $\nabla^2\rho(\mathbf{r}_{\mathrm{BCP}})$ values and the three-dimensional shape of the static deformation density (extension perpendicular to the bond). This is coherent with the integrated charges. The negative charge was found to be distributed over the whole ring instead of being localised at an anionic centre. This was further substantiated by qualitatively and quantitatively comparable lone-pair VSCCs at the two chemically different nitrogen atoms of the donor and the anion, respectively. The potential carbanion did not reveal the expected characteristics. In the Laplacian distribution no well-defined charge concentration could be detected, which allowed the identification of a lone-pair. Merely an asymmetrical distribution of the negative Laplacian on a relatively low level was observed. However, even if this could not be identified as a lone-pair, a bond induced polarisation of the valence shell of the carbon atom by the lithium cation is visible.

Again, it was the three-dimensional distribution of the ESP which was the key for the understanding of the observed reactivity. The only spatial region with negative ESP was found above the picolyl anion plane face-off the aza–allylic bonds. This region is extended towards the nucleophilic carbon atom of the methylene group. Here, the electrophilic attack is expected to occur. The three-dimensional shape fuels the idea that it is the π-density as a whole instead of a localised charge at a carbanion which determines the reactivity. The authors state that the ESP suggests "that potential electrophiles are literally guided by the negative potential towards the nucleophilic C6 atom" (Fig. 6).

This motif of the distribution of the ESP leading to the above mentioned reactivity was already observed in a previously published study on an α-lithiated benzylsilane [15]. There the question arose whether substitution of a highly diastereochemically enriched silyl-substituted alkyllithium follows the retentive pathway or causes an inversion. The investigation was undertaken at (*R*,*S*)-2•quinuclidine (**9**). As in **10**, the

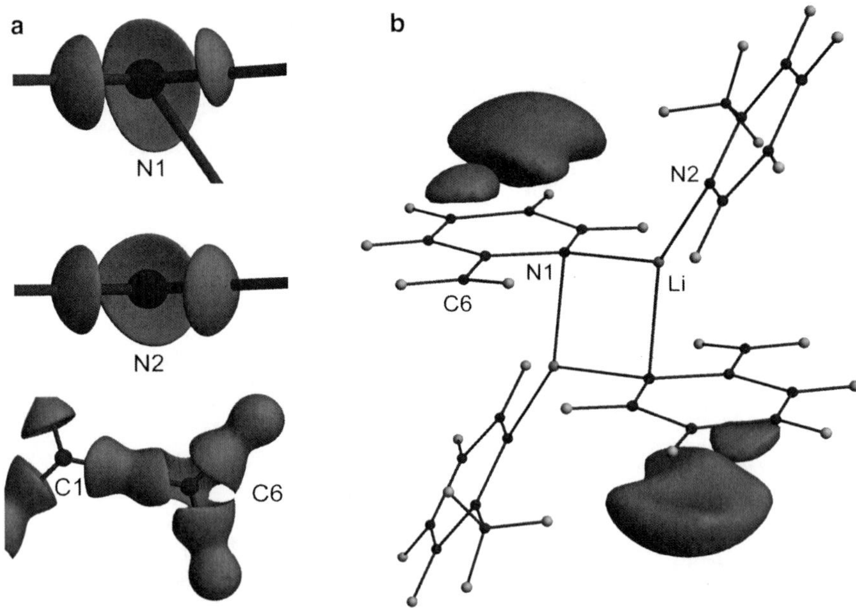

Fig. 6 $-L(\mathbf{r})$ (**a**) and ESP (**b**) in **10** (taken from [14])

perimeter bonded to the potential carbanion was in the focus of interest. In contrast to 2-picolyllithium no indicators for distinct charge delocalisation into the ring system were found as a consequence of the deprotonation. On the contrary, here the charge seemed to be localised at the potential carbanion. This was attributed to the integrated charges of almost -1 of the carbon atom and no distinct increase of density or negative Laplacian at the BCP. The calculation of integrated atomic charges was extended to so-called group charges. This allowed the identification of the anionic and cationic units. Interestingly, two cationic units were identified. In addition to the lithium cation, $SiMe_2$ was found to be positively charged with +1.22 e. The positive charge is counterbalanced by OMe (-0.39 e), the pyrrolidine (-1.08 e) and the benzyl group (-0.87 e).

The bonding characteristics were discussed in detail, leading to a fine-graded differentiation between the Li–E bonds within the molecule. All of them were identified to be severely polarised and dominated by ionic interactions. Li^+ was identified to be the origin of these findings. Lithium polarises all bonding partners but to a different degree. The Laplacian distribution along the respective bond paths gave a stringent picture. All Li–E bonds were characterised by charge concentrations exclusively located at the electronegative bonding partner and charge depletion over most of the interatomic region. In this context, it might be of interest that the Li–O bond showed the same unique distribution with a second minimum (on depletion level) mentioned above, which makes it easy to distinguish from all other so-called "dative" bonds.

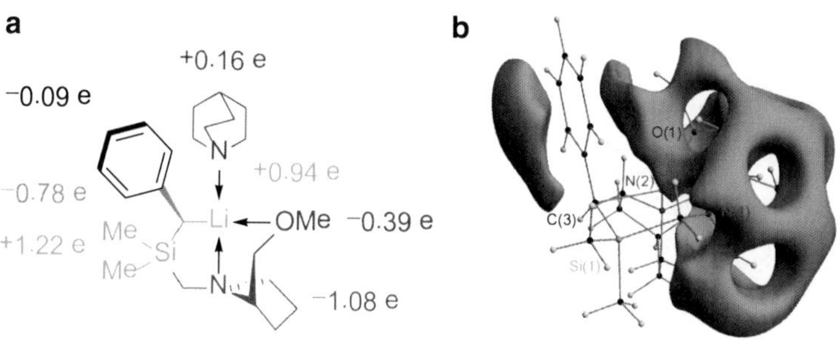

Fig. 7 Integrated group charges (**a**) and ESP (**b**) in **9** (taken from [15])

Even the above mentioned bifurcated donation towards Li^+ was observed. However, the authors could not resolve two lone-pair associated charge concentrations. They found them to be merged banana-shape like, but totally different to the well-defined lone-pairs at the nitrogen donors. At the carbanion no well-defined lone could be detected. Comparable to the findings in **10** the three-dimensional distribution of the negative Laplacian revealed no sharp maximum but a polarisation on low level. Nevertheless, the polarisation pattern is more distinct compared to **10**.

The question whether a possible electrophilic attack occurs from the front or back side and therefore the identification of the reaction mechanism via retention or inversion was again answered by inspection of the ESP. A comparable distribution to **10** was found and as there the reactant seems to be guided to the formal carbanion forcing a back-side attack with inversion (Fig. 7).

All published examples of this group of compounds (C^--type) so far have one thing in common. They seem to be paradigmatic examples where the chemist can get *predictive* information from the experimental charge density distribution. Here it is not the structural post-synthetic information, which is in the focus but first and foremost the deduction of properties which might allow to plan and predict properties instead of just understanding chemical behaviour.

3.3 *The Charge Density Distribution in Structures of the δLi-Type*

Thinking of lithium organics, the deltahedral lithium cores are for sure the most interesting species. However, little is known from the experimental charge density point of view up to now. Theoretical investigations dealt with the extremely interesting features of planar lithium clusters [3]. In these studies Gatti et al. investigated the potential interactions between the lithium cores. They found no direct bond path between the lithium atoms but indirect connection via so-called non-nuclear attractors, connecting the neighbouring atomic Li-basins. Lithium organic compounds by means of a topological analyses of the electron density distribution were for the first time theoretically investigated by Bader and MacDougall already in 1985 [4] when they deduced the reactivity of, e.g. $LiCH_3$

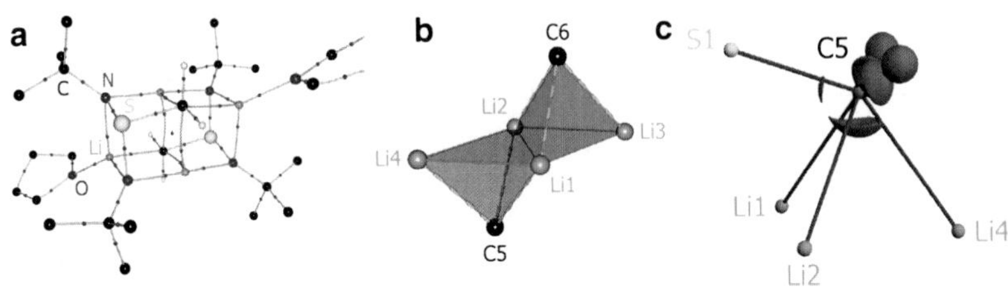

Fig. 8 Crystal structure made up by the bond paths (**a**), Li_3C motif (**b**) and distribution of the negative Laplacian at the carbanion in **7** (**c**) (taken from [12])

from the topological characterisation and then later by Ritchie and Bachrach [40]. The lithium–carbon bonds were inspected, e.g. for the methyllithium tetramer. They found bond paths between the carbanions and all three μ_3-capped lithium cations. Further investigations dealt with the application of QTAIM to lithium organics [41–43] but it took several years until Deuerlein et al. published the first experimental study on carbon-capped lithium triangles, the fundamental building block of deltahedral lithium cores, in 2008 [12].

In this molecule the structural fragment of a lithium triangle is embedded in a dimeric $S_2N_4C_2Li_4$ double cube, where the two cubes are fused through a Li_2C_2 face (see Fig. 8). In **7** it was not only the lithium–carbon interaction which was of interest, but also the S–N bonds were under debate, since they were another example of potential hypervalent sulphur–nitrogen double bonds. The topological properties of the S–N as well as the S–C bonds were straightforwardly characterised to be polar single bonds and therefore in line with earlier findings in comparable species. Dative lone-pair driven bonds between the nitrogen and the lithium atoms were characterised and the sulphur nitrogen interactions classified as being of the $S^+–N^-$ type instead of a double bond ruling out the hypervalent bonding model as sufficient explanation of the observed bonding modes. The integrated charges mirror the bonding properties with distinct negative values for the nitrogen atoms.

A new application of a very detailed analysis of the negative Laplacian distribution was presented there. The local maxima in the negative Laplacian distribution were determined for all non-hydrogen atoms in the molecule. These points in space refer to local charge concentrations of the respective valence shells due to bonding interaction with the neighbours or lone-pairs. The maxima were used to define an alternative geometry of the molecule by calculating angles between the atoms based on these critical points. This alternative geometry is electron density based and reflects the electronic state of a given molecule much better. As can be seen in Fig. 8, the central structural element is a double cube made up by $S_2N_4C_2Li_4$. Already on first sight it can be seen that the bond angles calculated from the atomic core positions cannot give much insight into the electronic state, e.g. the hybridisation of the involved atoms, since all angles in the cube are more or less close to 90°. The angles calculated from the maxima of the local charge concentrations fit the deduced hybridisation state extremely well and the predictions made from the VSEPR theory [44]. In this context, it was straightforward to identify the sulphur and nitrogen atoms to be sp^3-hybridised with well-defined lone-pairs face-off the cube at the sulphur atoms and lithium-directed at the nitrogen atoms. The cube-defining bonds between the sulphur and nitrogen atoms as well as the sulphur–carbon bonds were identified as charge assisted single bonds instead of double bonds which were suggested from the concept of hypervalency.

However, undoubtedly of major interest for the synthetic chemist, it was neither the S–N nor S–C bonds which were in the focus in **7**. It is the μ_3 carbon-capped lithium triangle, which was accessible for the first time for a topological analysis based on experimental data. An extremely well-resolved lone-pair related charge density concentration which is directed towards the lithium triangle was found at the carbon atom. Interestingly, the orientation of this lone-pair relative to the Li_3 triangle is not symmetrical. It has a preferred orientation towards the lithium atom at the tip of the isosceles triangle made up by Li_3. Nevertheless, BPs were found to all three lithium atoms with relatively low values for $\rho(\mathbf{r}_{BCP})$ and positive $\nabla^2\rho(\mathbf{r}_{BCP})$. The existence of these BPs has to be taken as the sufficient condition that the carbanion and all capped lithium anions are bonded to one another [45]. Therefore, the authors conclude that the carbanion is sp^3 hybridised which can be deduced from the number and orientation of the four VSCCs, on the other hand if the coordination mode is of interest, the carbon atoms in the double cube should be understood as sixfold coordinated: 2 covalent single bonds to the hydrogen atoms, a charge assisted (S^+–C^-) single bond to the sulphur atom at the edge of the cube, and three closed shell bonds to the lithium cations of the capped triangle.

The latest example of a deltahedral lithium core, the trimethylsilylmethyllithium hexamer (**8**) was presented by Ott in 2009 [13]. The structure is still under investigation but the main structural features are given and they complete the picture already presented for the lithium sulphur ylide. Six lithium cations form a distorted octahedron in the hexamer. Six of the eight isosceles triangular faces made up by three lithium atoms are μ_3-capped by $SiMe_3H_2C^-$ anions. Like in **7**, BPs between the carbanion and all three capped lithium atoms are found. In contrast to **7**, the orientation of the lone-pair gives no hint to a preferred donation of the lithium at the tip of the isosceles triangle but its projection to the Li_3 face is closer to the base. Interestingly, this fits to the topological features of the bond path analysis. On very low level but stringent through the whole molecule, the Laplacian distribution along the BP is to distinguish for the three independent C^-–Li^+ bonds of each triangle.

The analysis of the VSCCS around the carbanions gave four local maxima in the negative Laplacian, three oriented towards the hydrogen atoms and the silicon neighbour and the one of the lone-pair directed towards the Li_3 face (Fig. 9). The carbon atoms are therefore sp^3-hybridised, which can be deduced from the number of VSCCs and the related angles calculated from the local maxima in the negative Laplacian and, as in **7**, they are sixfold coordinated. The lone-pair charge concentrations are in the mean even higher (-21.5 to -28.3 e/Å^5) compared to the carbanions in the sulphur ylide (-23 e/Å^5) but significantly smaller compared to the *t*Buli·(–)-sparteine (-34 e/Å^5), which can be attributed to the inductive effect of the methyl substitution and the coordination to one single lithium cation. The fine graded but coherent differences in the quantities of the VSCCs might be

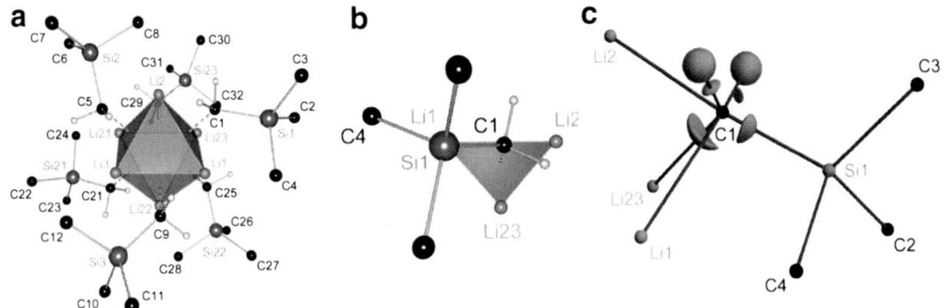

Fig. 9 Molecular structure (**a**), Li_3C motif (**b**) and distribution of the negative Laplacian (isolevel $-15\ e/Å^{-5}$) (**c**) in **7** (taken from [13])

a measure for "dative potential" or "potential reactivity" of donor atoms in future works.

The integrated charges fit into the picture given by the Laplacian distribution and the bond topology. The mean charge of the lithium ions is +0.85 e and the carbanions exhibit a surprisingly high negative charge of −1.57 e each. The bonding between the CH_2 groups and the positive lithium cores is therefore doubtlessly charge-controlled leading to the interpretation of the C^-–Li_3 interaction as a four centre-two electron bond (4c-2e) following the nomenclature in borane chemistry.

Concerning the potential Li–Li interaction the findings are also unambiguous. No BPs are found between the lithium atoms, which do not share a common zero flux surface. This makes the propagation of a BP impossible. Bonding in the topological sense can therefore be excluded.

However, a secondary effect could be observed, by inspection of the bonding in the CH_2SiMe_3 units. Significant differences were detected for the three methyl groups depending on their orientation towards the lithium core. The methyl carbon atoms pointing towards the lone-pair of the carbanion reveal a reduced charge compared to the *trans* oriented methyl carbon atom. This is accompanied by ellipticities along the BP which show the same characteristics as already observed by Scherer et al. in their alkyllithium complex, **1** [6]. The interpretation is therefore also in line: negative hyperconjugation is believed to stabilise the negative charge at the deprotonated carbon atom in the hexamer, leading in consequence to a shortening of the methylene–silicon bonds and elongation of the *trans* oriented silicon–methyl bond. The Si–C_α bonds show partial multiple bond character due to a presumed interaction of the lone-pair density with the σ^*-orbital of the bond between silicon and the *trans* oriented methyl carbon atom.

4 Perspective

Even though just a few examples for topological analyses of the experimental electron density are known, the molecules published so far represent a wide spectrum of lithium organics and the results gained until now lead to a more comprehensive picture of the electronic situation in this class of compounds.

The findings for the lithium organics listed here lead to results reaching much further than standard structural information or even "standard" topological analysis. Based on the typical procedures like bond path analysis, calculation of the two- and three-dimensional Laplacian distribution and its critical points, calculation of integrated charges, deep rooted knowledge on these compounds could be obtained.

New interpretations of ambigous interactions between atoms can be given, like the differentiation between so-called hypervalent bonding and charge-assisted interaction, the proof for the non-existence of Li–Li bonds in the deltahedral lithium metal cores, the direct evaluation of negative hyperconjugation, and first (experimental) insight into the interaction of the carbanions μ_3-capping the Li_3 triangles, which is the fundamental building block for a huge class of compounds.

On the other hand, the studies led to the recommendation of an "alternative geometry" based on the angles calculated from the position of local maxima in the negative Laplacian. These density-based values were in line with the predictions from VSEPR theory when the calculation from the core positions in non-standard environment failed, like it is realised in the deltahedral lithium organics.

Since all these "interpretive" statements are of course susceptible for controversial interpretation, it is of fundamental importance that the predictive power of the method of topological analysis on experimental data had been shown by the structure-reactivity studies on the octamethylcyclotetrasilazane and its deprotonated lithium complexes, on 2-picolyllithium, and on an α-lithiated benzylsilane. In all these studies the reactivity of the species could be explained and/or predicted by inspection of the electrostatic potential calculated from the experimental charge density. A test on the bases of experimentally derived *predictions* should be the ultimate proof for any scientific approach. In this sense, the topological analysis of experimentally derived charge density distributions seems to be a promising technique on the way to a reliable structure–property relation.

Acknowledgement The author gratefully thanks Dr. Holger Ott for his careful proof reading, helpful comments and scientific input to this chapter.

References

1. Coppens P (1997) X-Ray charge densities and chemical bonding. Oxford University Press, Oxford
2. Bader RFW (1990) Atoms in molecules - a quantum theory. Oxford University Press, New York
3. Gatti C, Fantucci P, Pacchioni G (1987) Theor Chim Acta 72:433–458
4. Bader RFW, MacDougall PJ (1985) J Am Chem Soc 107:6788–6795
5. Reed AE, Weinstock RB, Weinhold F (1985) J Chem Phys 83:735–746
6. Scherer W, Sirsch P, Grosche M, Spiegler M, Mason SA, Gardiner MG (2001) Chem Commun 2072–2073
7. Scherer W, Sirsch P, Shorokhov D, McGrady GS, Mason SA, Gardiner MG (2002) Chem Eur J 8:2324–2334

8. Stey T, Stalke D (2004) In: Rappoport Z, Marek I (eds) The chemistry of organolithium compounds. Wiley, Chichester, pp 47–120
9. Stalke D (1998) Chem Soc Rev 27:171–178
10. Kocher N, Leusser D, Murso A, Stalke D (2004) Chem Eur J 10:3622–3631
11. Kocher N, Selinka C, Leusser D, Kost D, Kalikhman I, Stalke D (2004) Z Anorg Allg Chem 630:1777–1793
12. Deuerlein S, Leusser D, Flierler U, Ott H, Stalke D (2008) Organometallics 27:2306–2315
13. Ott H (2009) Dissertation, University of Goettingen, Germany
14. Ott H, Pieper U, Leusser D, Flierler U, Henn J, Stalke D (2009) Angew Chem 121:3022–3026; Angew Chem Int Ed 48:2978
15. Ott H, Däschlein C, Leusser D, Schildbach D, Seibel T, Stalke D, Strohmann C (2008) J Am Chem Soc 130:11901–11911
16. Storm AB, Oehr CMA, Hoffmann C (2004) Proc SPIE 5537:177–181
17. Coles SJ, Hursthouse MB (2004) J Appl Crystallogr 37:988–992
18. Schulz T, Meindl K, Leusser D, Stern D, Graf J, Michaelsen C, Ruf M, Sheldrick GM, Stalke D (2009) J Appl Crystallogr 42:885–891
19. Leusser D (2002) Dissertation, University of Wuerzburg, Germany
20. Bruker AXS (2006) APEX2, Madison, WI, USA
21. Farrugia LJ (1999) J Appl Crystallogr 32:837
22. Bruker AXS (2002) SAINT, Madison, WI, USA
23. Kabsch W (2010) Acta Crystallogr D66:125–132
24. Duisenberg AJM, Kroon-Batenburg LMJ, Schreurs AMM (2003) J Appl Crystallogr 36:220–229
25. Sheldrick GM (2002) SADABS version 2.03, University of Goettingen, Germany
26. Blessing RH (1995) Acta Crystallogr A51:33–37
27. Zhurov VV, Zhurova EA, Pinkerton AA (2008) J Appl Crystallogr 41:340–349
28. Hansen NK, Coppens P (1978) Acta Crystallogr A34:909–921
29. Volkov A, Macchi P, Farrugia LJ, Gatti C, Mallinson PR, Richter T, Koritsanszky T (2006) XD2006, a computer program package for multipole refinement, topological analysis of charge densities and evaluation of intermolecular energies from experimental or theoretical structure factors
30. Allen FH (1986) Acta Crystallogr Sect B 42:515–522
31. Sheldrick GM (2008) Acta Crystallogr Sect A 64:112–122
32. Madsen AØ (2006) J Appl Crystallogr 39:757–758
33. Dittrich B, Koritsanszky T, Luger P (2004) Angew Chem 116:2773–2776; Angew Chem Int Ed 43:2718
34. Kirzhnits DA (1957) Sov Phs - JEPT 32:115
35. Cremer D, Kraka E (1984) Angew Chem 96:612–614; Angew Chem Int Ed Engl 23:627
36. Kocher N (2003) Dissertation, University of Wüerzburg, Germany
37. Mebs S, Grabowski S, Förster D, Kickbusch R, Hartl M, Daemen LL, Morgenroth W, Luger P, Paulus B, Lentz D (2010) J Phys Chem A 114:10185–10196
38. Mebs S, Kalinowski R, Grabowsky S, Förster D, Kickbusch R, Justus E, Morgenroth W, Paulmann C, Luger P, Gabel D, Lentz D (2011) J Phys Chem A 115: 1385–1395
39. Macchi P (2009) Angew Chem Int Ed Engl 48:5793–5795
40. Ritchie JP, Bachrach SM (1987) J Am Chem Soc 109:5909–5916
41. Ponec R, Roithova J, Gironés X, Lain L, Torre A, Bochicchio R (2002) J Phys Chem A106:1019–1026
42. Grotjahn DB, Pesch TC, Xin J, Ziurys LM (1997) J Am Chem Soc 119:12368–12369
43. Cooper DL, Gerratt J, Karadakov PB, Raimondi M (1995) J Chem Soc Faraday Trans 91:3363–3365
44. Gillespie RJ, Popelier PLA (eds) (2001) Oxford University Press, New York
45. Bader RFW (2009) J Phys Chem A 113:10391–10396

Struct Bond (2012) 146: 127–158
DOI:10.1007/430_2010_30
© Springer-Verlag Berlin Heidelberg 2010
Published online: 27 November 2010

Bond Orders in Metal–Metal Interactions Through Electron Density Analysis

Louis J. Farrugia and Piero Macchi

Abstract The metal–metal bond is central in the chemistry of polymetallic complexes. Many structural investigations, both experimental and theoretical, have been carried out with the purpose of understanding this interaction in more detail and of being able to predict the stereochemistry of these molecules. Among these studies, increasing importance is given to electron density analysis. Originally, only deformation densities were analysed, but it became clear that more sophisticated theories were necessary to appreciate the subtleties of these elusive chemical bonds. Thus, the quantum theory of atoms in molecules, electron delocalisation indices, the electron localisation function and the domain averaged Fermi hole density are nowadays used to characterise metal–metal bonds. The major results reported in the literature in the past few years are carefully reviewed in this chapter.

Keywords Electron density · Atoms in molecules · Bond order · Metal-metal bond · Transition metal complexes

Contents

L.J. Farrugia (✉)
Department of Chemistry, University of Glasgow, Glasgow G12 8QQ Scotland, UK
e-mail: louis.farrugia@glasgow.ac.uk

P. Macchi (✉)
Department of Chemistry and Biochemistry, University of Bern, Freiestrasse 3, CH3012 Bern Switzerland

1 Introduction

The Lewis model [1] of covalent bonding between atoms, despite being proposed nearly 100 years ago, is still a central idea in chemistry, even though it is quite inapplicable in certain areas such as weak intra- and intermolecular interactions or metallic bonding in periodic solids. The quantum theory of atoms in molecules (QTAIM) [2], on the other hand, provides an alternative description of chemical bonding, without suffering from the limitations of the Lewis model. Nevertheless, the fundamental and continuing importance of the Lewis model to chemical thinking has driven a search for a concordance between these two differing approaches, see for example [3–5]. In this chapter, we examine, through QTAIM analysis of the electron density, the nature of the metal–metal (M–M) interaction in *molecular* compounds containing formal M–M bonds, or otherwise in the absence of any formal bond. The examples we discuss involve primarily, though not exclusively, the *d*-transition metals. The original Cotton definition [6] of a cluster compound as a "group of metal atoms held together mainly or at least to a significant extent, by bonds directly between metal atoms" has now been significantly diluted by the current habit of referring to all polynuclear metallic complexes (which often possess no formal metal–metal bonds between the metallic centres) as "clusters". As this chapter is concerned with the nature of the metal–metal interaction in molecular compounds, we will hold to the original Cotton definition [6] when using the term "cluster compound". It must be stressed, however, that there is considerable fluidity in the concept of a metal–metal interaction. For example, there are many cases where there is clear evidence for strong magnetic exchange between metal ions in complexes (though with no formal metal–metal bonds), such as in the very topical area of single-molecule magnets [7]. Furthermore, the relationship between the concept of metallic bonding in extended solids and metal–metal bonding in discrete molecular compounds is not a clear one, as discussed below. Finally, as will be stressed throughout this chapter, the definition of direct M–M bonding is actually quite controversial and QTAIM analyses of the electron density has revealed numerous discrepancies, even in cases where simple "chemical intuition" would seem to provide a clear answer.

Within the terms of the original Cotton definition of a molecular cluster compound, we can identify three general classes of metal–metal bonded species: (1) organometallic dimers, clusters and chains containing π-acid ligands, especially the ubiquitous carbonyl ligand; (2) complexes (mainly dimetallic) with formal multiple metal–metal bonds – examples here are found for organometallic and classic coordination species; and (3) high oxidation state metal clusters with π-donor ligands (e.g. oxo, sulfido, halides). The metal atoms in these clusters are almost invariably drawn from the *d*-transition series, and for the vast majority of examples from class (1), the formal M–M bond order is one. The general features of the chemical bonding in all these systems have been extensively studied over the last 40 years, using the orbital approach in various levels of approximation. These studies have allowed the expansion of the simple 18-electron or effective atomic number (EAN) rule, which has a somewhat limited predictive utility in the case of small metal clusters only, to the general Wade/Mingos rules governing the relationship between electronic structure and metal-core geometry [8, 9]. The delocalised nature of the M–M bond in larger clusters has been clearly demonstrated, and the relationship between M–M interactions, formal bond orders and individual 2c–2e M–M chemical bonds has been shown to be quite ambiguous (often invalidating the counting schemes).

The chemical bonding between metal atoms in organometallic molecules is obviously different from that in metal conductors, although the electron density distribution may not be able to clearly differentiate the two. The *metallic* bond has been difficult to include in bonding classifications. Pauling originally suggested [10] it should be considered as a partial covalent bond. In agreement with this view, Anderson et al. [11] and Allen and Capitani [12] have suggested it should be removed from any special bonding classification altogether, and treated instead as a special case of covalency. More recently, Silvi and Gatti [13] have carefully discussed the possibility of describing the *metallic* bond in direct space. This work implies that the electronic conduction could be revealed by mapping a quantum mechanical observable to elastic scattering experiments, instead of spectroscopic methods. The conclusion of Silvi and Gatti [13] that the metallic bond is basically a partial covalent bond (partial in the sense that the electron population associated with the bonds is low) is in line with earlier conclusions. Despite their study, little further work has been carried out in this direction. Moreover, it should be noticed that less and less theoretical studies are dedicated to the metallic bond, the interpretation of which is still based on old concepts and very briefly described in textbooks. One may admit here the inherent difficulty of retrieving genuine information on the electronic conduction from the electron density. For this reason, caution should also be exercised when studying bonds between metals in discrete molecular compounds. The distributions of electrons around metals are dominated by the core and inner valence, whereas the outer valence (typically the *ns* electrons) is associated with very diffuse electron density, which is difficult to visualise. Sometimes, small clusters of metal atoms (or even the infinite crystalline state) give rise to electron density maxima at non-nuclear positions [14]. These are the non-nuclear attractors (NNAs) whose characterisation and interpretation have been subject of several discussions in the literature.

The need, therefore, for supporting experimental evidence as to the nature of these unusual chemical bonds was quickly recognised. In fact, many of the early experimental charge density studies on transition metal compounds focused on M–M bonding, with examples including $Mn_2(CO)_{10}$ [15], $Co_2(CO)_8$ [16], *trans* $(\eta^5\text{–}C_5H_5)_2Fe_2(CO)_4$ [17], $(\eta^5\text{–}C_5H_5)_2Ni_2(\mu\text{–}\eta^2\text{–}C_2H_2)$ [18], $(\eta^5\text{–}C_5H_5)_2Mn_2(CO)_4(\mu\text{–}CH_2)$ [19], $(\eta^5\text{–}C_5H_5)_2Cr(\mu\text{–}\eta^8\text{–}C_8H_8)$ [20] and $Cr_2(O_2CCH_3)_4(OH_2)_2$ [21]. All these experimental studies involved the examination of rather noisy deformation density maps, and despite the formal M–M bond orders of at least 1 for all complexes, no *significant* charge density build ups at the M–M bond centres were observed (Fig. 1). Theoretical deformation maps showed similar results, and this was originally attributed to the "diffuse character of the metal *d*-orbitals, which favours large accumulation regions of low density gradient,

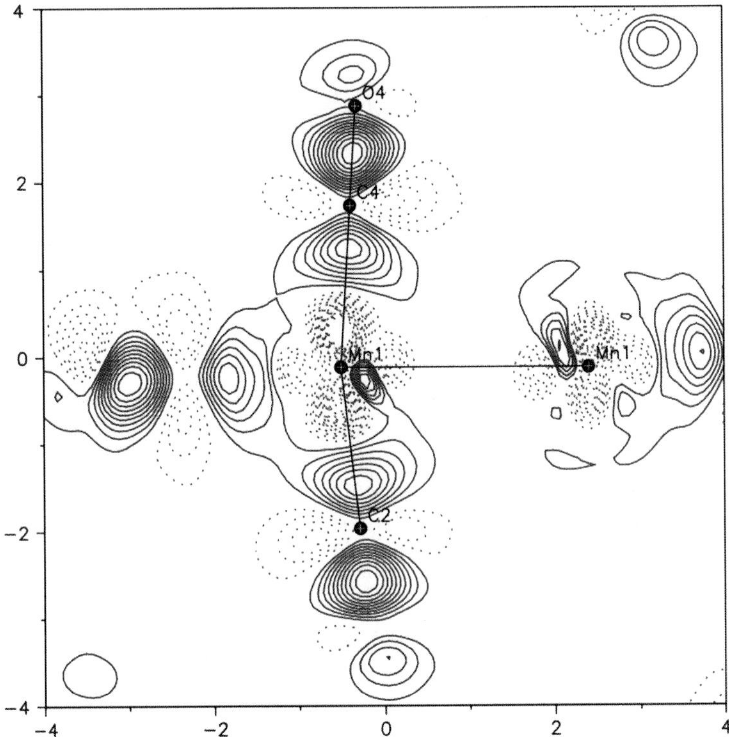

Fig. 1 Dynamic model deformation map for $Mn_2(CO)_{10}$, showing the virtually flat density in the region of the Mn–Mn midpoint. Contours at 0.05 eÅ^{-3}, positive contours as *solid lines*, negative contours as *broken lines*

rather than sharp peaks in limited regions" [21]. It should be noted, however, that pure *d*-orbitals are rather contracted and the real reason for the missing peaks is due to low overlap. The argument in [21], however, holds true for *s*-orbitals and for *sd*-hybrids.

Anyway, no clear evidence for the covalent nature of M–M bonds was forthcoming, and these rather disappointing initial results inevitably led to some scepticism as to the usefulness of charge density analyses in this area. Although more accurate data might have provided a more confident result, the nature of the problem was not fundamentally one of the experimental error, but rather methodological. For this reason, the QTAIM approach [2] was used later in attempting to overcome interpretative problems. An early topological analysis of the theoretical electron density in $Co_2(CO)_8$ by Hall and co-workers [22] showed definitively the lack of a bond critical point (BCP) between the two Co atoms at the equilibrium geometry, indicating there to be no *direct* Co–Co bonding (whereas a bent bond was suggested on the basis of deformation density maps). Interestingly, the density in the Co–Co region was shown to be very flat and the topology quite sensitive to the Co–Co distance, thus providing an initial indication of the catastrophic nature [2] of the M–M interaction. On the other hand, in $Mn_2(CO)_{10}$ an Mn–Mn bond path was found [23]. Thus, species with similar formal bond orders appeared with different molecular graphs and topologies.

It soon became clear, however, that a straightforward application of QTAIM "rules" (derived from the electron density distribution of organic molecules) would not be particularly appropriate for M–M bonded species. The peculiar nature of metal atoms makes the electron density distribution in their compounds quite different. In particular, we can identify two important questions: (1) what is the meaning of a bond path in molecules where the Lewis bonding model is not working? (2) what is the nature of chemical bonds involving metals? The first question is tantamount to asking whether we should assign the presence of a two-centre–two-electron bond when a bond path between two atoms is found? The answer is resoundingly in the negative, as repeatedly stressed by Bader [24, 25]. Nothing in the bond path derivation tells us about any such relationship, rather the bond path can be associated with the *virial* of forces linking atoms together, through the structural homeomorphism of the two fields [26]. As a note of caution, it should be stressed that the homeomorphism between the molecular graphs derived from electron density and those from potential energy density is not perfect, particularly in regions of very flat density, where a catastrophe situation may arise [26, 27] – this caveat of course being especially pertinent to M–M bonds. Recent work by Pendás et al. [28] has suggested an alternative interpretation of the bond path in terms of the preferred pathway of quantum mechanical exchange, but this approach also provides no link between the bond path and a Lewis model of a two-centre–two-electron bond. The answer to the second question is still an open debate among scientists, and we will try to elucidate it in this chapter. Indeed, the amount of electron sharing and the degree of delocalisation in metal–metal bonds are the central points. A genuine distinction from other types of chemical bond is not always so easy, because the small electron concentration within an M–M bond is a feature expected also for pure closed-shell interactions (like van der Waals interactions).

In terms of the Lewis model, and in the minds of practising chemists, the formal bond order is an important concept. Unfortunately, this concept is not a quantum mechanical observable, and it is not carried over in any formal sense into the QTAIM domain, where a continuum of descriptors is normally applicable. Various empirical relationships between QTAIM indicators such as ρ_{BCP} and a bond order such as (1) have been proposed [2], where A and B are constants, dependent on the bonded atoms.

$$BO = \exp\left[A(\rho_{BCP} - B)\right]. \tag{1}$$

Such definitions, however, lack rigour and are not generally applicable. The delocalisation index $\delta(\Omega_A, \Omega_B)$ between two atomic basins A and B, originally proposed by Bader and Stephens [4], has a clear interpretation at the HF level as the number of pairs of electrons shared between two basins, and so has an obvious link to the bond order in the Lewis sense. Unfortunately, this index requires in the general case information about the first and second order density matrices, and so is not obtainable solely from an experimental electron density distribution (at least, as far as the current state of affairs concerned) [29].[1] Moreover, the interpretation is less clear when the investigated bond is quite polarised or highly delocalised. For the purpose of this chapter, we apply the following working classifications of M–M bonding:

1. Through-bridge bonding, where any direct M–M interaction can be reasonably excluded
2. Partial direct M–M bonding, where a direct interaction is evident or otherwise cannot be deconvoluted from through-bridge bonding
3. Multiple bonding, where a direct M–M interaction can be further decomposed in different contributions

One particularly illustrative example of the difficulties in reconciling "intuitive" chemical views of M–M bonding with the modern QTAIM approach is provided by the classic molecule $Fe_2(CO)_9$. This was the first metal carbonyl compound whose structure was determined by X-ray crystallography [30], and a later and more accurate determination by Cotton and Troup [31] revealed the D_{3h} molecular symmetry, with three symmetrically bridging CO groups and a short Fe–Fe distance of 2.523(1) Å. Application of the EAN rule leads to the conclusion of a single direct Fe–Fe bond. The short internuclear distance, coupled with the diamagnetic nature, is quite consistent with this idea – a view in fact promulgated for many years in inorganic text books (see, e.g. [32]). Nevertheless, numerous MO studies over the years [33–37] have found little evidence for any direct Fe–Fe bonding, though some authors [38] claim that a small attractive direct Fe–Fe interaction is present. An early QTAIM analysis by Bo et al. [37] found only a minimum in the charge density at the Fe–Fe midpoint (i.e. a cage critical point), leading to the conclusion that no direct Fe–Fe bonding was present. However, a later QTAIM study by Reinhold et al. [39] was more ambiguous and showed that the nature of the critical point at the Fe–Fe midpoint was highly dependent on the basis set and in some cases a BCP could be observed. Using an orbital partitioning of the total density, these same authors [39] concluded that some weak direct Fe–Fe bonding was present. This viewpoint was further emphasised, in their opinion, by the relatively large delocalisation index $\delta(\Omega_{Fe}, \Omega_{Fe})$ of 0.4. On the other hand, using another technique for the real-space analysis of the wavefunction, the domain averaged Fermi Hole (DAFH; defined over QTAIM atomic basins) [40], Ponec and co-workers [41, 42] provided clear evidence which directly contradicts this view. These studies confirmed the 3c–2e nature of the Fe–Fe bonding through participation of the bridging carbonyl ligands, but provided no evidence for any direct Fe–Fe bonding. In summary then, the totality of evidence to date strongly favours a delocalised 3c–2e carbonyl-bridged bonding view of the Fe–Fe interaction, rather than any direct bonding – a view now finding

[1] An X-ray constrained wave function approach [29] obviously would make the delocalisation index available, though lacking any strict physical meaning.

favour in some modern text books [43]. In most cases, the density in the M–M bonding region is very flat, as seen above for $Co_2(CO)_8$ and $Fe_2(CO)_9$, which can lead to ambiguities in the topology, i.e. the bifurcation catastrophe situation [2]. It follows that any connection between the topology of the electron density and electron counting schemes such as the Mingos/Wales rules [8, 9] is not possible. In this area at least, it should be emphasised that QTAIM has no predictive power.

In the remainder of this chapter, we will discuss the problems of metal–metal bonding by considering the *formal* bond order. To focus the readers' attention, it is worthwhile emphasising here our main conclusions, which are further developed in the course of our discussions.

1. The presence of an M–M bond path, indicative of chemical *bonding*, as stressed by Bader [24, 25], clearly implies some M–M interaction. However, due to the general diffuse nature of M–M bonds, the usual criteria to define the strength of such an interaction, e.g. ρ_{BCP}, are not applicable. The density integrated over the mutual interatomic surfaces is a better indicator in this case.
2. It is an empirical observation that the presence of any bridging ligand will result in the loss of the M–M bond path, at least for formal single M–M bonds. For formal multiple M–M bonds, this is no longer the case.
3. Other important criteria for defining the presence (or otherwise) and nature of an M–M interaction are the delocalisation index, and more visually the DAFH.
4. Simple electron counting schemes like the EAN or the more complex Wade/Mingos rules [8, 9], while useful in categorising chemical compounds, do not give great insight into the nature of any M–M bonding.
5. There is generally a lack of correlation between the QTAIM indicators, especially those computed at just one point, and "chemical" indicators such as bond length, bond strength and particularly bond order. This is especially noticeable in the case of multiple M–M bonds.

2 Bond Order = 0: Through-Bond Interactions

Ligand bridges between two metals are very common and they often provide more stable complexes and more robust architectures. Considerable discussion has been dedicated to distinguishing between direct M–M bonds and indirect (through-bond, or even non-bonding) M–M "contacts". Any scheme suffers from ambiguity and a universally accepted interpretation is not yet available. In terms of analysis of the electron density, there is clearly a problem. For example, while the presence of a deformation density peak in the middle of an interatomic vector would usually be taken as a clear indicator of a direct bonding interaction, any delocalised through-bond interactions would not be detectable in this manner. Considering QTAIM criteria, on the other hand, one could question whether the presence of a bond path alone is a sufficient condition to imply *substantial* direct M–M bonding, i.e. is the presence of a base pair necessarily a significant observation regarding bond strength? It is well known that bond paths are often observed between atoms not formally sharing significant amount of electrons, for instance atoms involved in weak intermolecular interactions in crystals. Nor it is a necessary condition, as direct bonding is sometimes found in the absence of a bond path, see below. So it may be necessary to look beyond the molecular graph. The experimental and theoretical data for complexes having no critical point in the density associated with the M–M interaction are summarised in Table 1.

One interesting bridging ligand is the hydride. From an electronic point of view, M–H–M systems are interesting because they are a sort of "reverse" hydrogen bond, where the hydrogen is typically an electron donor and the metals are electron acceptors. There are very few experimental studies on these species. Apart from the obvious problems of detecting H atoms from X-ray diffraction experiments, the bridging hydride is often disordered over two positions, rendering experimental studies more complicated. The first full study on a typical M–H–M system was carried out on $[Cr_2(\mu_2\text{–}H)(CO)_{10}]^-$ [44], an almost symmetric system investigated as a salt with many different cations. The experimental data (for electron density purposes) were collected on the K^+ salt, but a theoretical investigation was performed on the full conformation space, spanning geometries actually observed in

Table 1 Topological properties of M–M formal single bonds without any associated critical points[a]

Compound	Refs.	M–M[b]	δ(M–M)[c]
$[FeCo(CO)_8]^-$	[79]	2.705	0.299
		2.6120(2)	–
$[Cr_2(\mu\text{–}H)(CO)_{10}]^{-d}$	[44]	3.340	0.095
		3.528	0.074
		3.567	0.067
		3.300(4)	–
$Co_2(CO)_6(\mu\text{–}\eta^2\text{–}CH{\equiv}CH)^e$	[131]	2.4898	0.536
		2.4685	0.261
$Co_2(CO)_6\text{–}(\mu\text{–}\eta^2\text{–}HC{\equiv}CC_6H_{10}OH)$	[83]	*2.465(1)*	–
$Co_4(CO)_{12}(\mu\text{–}\eta^4\text{–}PhC{\equiv}C\text{–}C{\equiv}CPh)^f$	[84]	*2.4584(1)*	–
$(\eta^5\text{–}MeC_5H_4)(CO)_2Mn[\eta^2\text{–}O{=}C{=}C((\mu\text{–}\eta^2\text{–}C{\equiv}CPh)Co_2(CO)_6Ph]$	[132]	*2.4696(1)*	–
$Co_3(\mu\text{–}CH)(CO)_9^f$	[80]	2.502	0.466
		2.479(1)	–
$Co_3(\mu\text{–}CCl)(CO)_9^f$	[75]	2.499	0.466
		2.477(1)	–
$Mn_2Cp_2(CO)_4(\mu\text{–}BR)$ (R=Me)	[81]	2.865	0.344
(R = ^{t}Bu)		*2.78190(7)*	–

[a]Theoretical values (if reported) shown on first line in plain font, with experimental results in italic font. Unless otherwise stated, calculations involve DFT optimised geometries using idealised symmetry at the B3LYP/6-311++G**/Wachters+f level
[b]Metal–metal distance, Å
[c]Delocalisation index between metal atoms $\delta(\Omega_M\text{–}\Omega_M)$
[d]First line optimised C_s geometry, second line optimised D_{4d}, third line optimised D_{4h} geometry
[e]Second line, calculation at CASSCF[6,6] level
[f]Co–Co distances averaged

other salts, as well as hypothetical ones. Complexes such as $[Cr_2(\mu_2\text{–}H)(CO)_{10}]^-$ were traditionally classified on the basis of the *open* or *closo* 3-centre bonding (see scheme 1). Stereochemical considerations led to the conclusion that a *closo* system was closer to reality, because the direction opposed to the axial ligands usually point towards the M–H–M ring centre rather than to the hydride, suggesting the presence of some direct M–M bonding. Notably, this hypothesis was formulated *before* the agostic interaction was actually discovered.

A QTAIM analysis, however, provides clear indications that the *closo* stereochemistry is not correct. On the one hand, there is no Cr–Cr bond path and, on the other hand, only minor differences in the Cr–H bond paths are observed when the linear geometry (necessarily implying no Cr–Cr direct interaction) and the bent geometries are compared theoretically. The analysis of the Fermi hole density showed that the Cr–Cr interaction is only a *through-bond* one, with no evidence of any direct coupling. This is also confirmed by the small values (0.104–0.088) of the delocalisation indices $\delta(\Omega_{Cr}, \Omega_{Cr})$, which increase only slightly as the system goes from linear to bent and which therefore indicate insignificant Cr–Cr electron sharing. The experimental electron density, although being in agreement with the theoretical prediction, is not able to fully prove this result. However, analysis of the Laplacian distribution of the hydride illustrates the similarity of this species with 3-centre-4-electron systems, lacking a direct pairing between the two external atoms. An unexpected role is played by the proximal equatorial carbonyl ligands, as visualised especially by the Fermi hole density distribution and the corresponding $\delta(\Omega_H, \Omega_C)$ delocalisation indices (δ about 0.1, see [44]). In addition, there is an electrostatic attraction between the negatively charged H atoms and the positively charged C atoms. This explains the actually observed stereochemistry, and definitely excludes any role for a direct Cr–Cr interaction on the observed geometries, despite a formal bond order of one from a consideration of the EAN rule.

Another class of compounds where some interaction between metal atoms could be suspected, but which have no *formal* metal–metal bonding, are those coordination complexes and metallo-oligomers with strong magnetic interactions between the metal ions. A number of these have been studied by QTAIM methods, by the groups of Iversen and co-workers [45–51], Lecomte and co-workers [52–54] and others [55]. In many cases, the large size of these molecules and their poor crystallinity has necessitated the use of synchrotron radiation to obtain sufficiently accurate experimental structure factors [56]. With one exception [47], metal–metal bond paths were not observed in any of these studies, and it was assumed that the

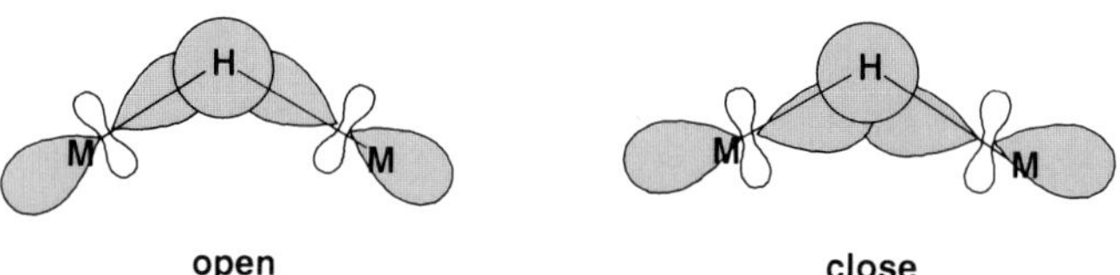

Scheme 1 *Open* or *closo* stereochemistry in metal–metal hydride dimers

magnetic interactions are of the super-exchange type, through the bridging ligands (primarily oxo groups). In the exceptional case of $[Mn_2(C_8H_4O_4)_2(C_3H_7NO)_2]_\infty$ [47], the density at the Mn–Mn BCPs is extremely low, and the significance of the Mn–Mn bond paths is questioned by the authors. Due to the lack of a BCP in these examples, little further investigations into the nature of any possible metal–metal interaction were undertaken. However, in the case of the dinuclear Cu(II) complexes $[Cu_2(ap)_2(L)_2]$ (ap = 3-aminopropanolate, L = NO_2^-, NO_3^-, HCO_2^-), Farrugia and co-workers [57] have confirmed the lack of any direct Cu–Cu bonding, since the delocalisation index $\delta(\Omega_{Cu}, \Omega_{Cu})$ of 0.02 is insignificantly small. A partial covalent character was demonstrated for the Cu–O bonds in the $Cu_2(\mu\text{–}O)_2$ unit, and so a super-exchange mechanism was assumed to account for the observed magnetic coupling. In this context, it is of interest to note the QTAIM studies of Jauch and Reehuis [58–60] on the solid oxides MO (M = Mn, Co, Ni), obtained using highly accurate experimental densities from γ-ray diffraction. They conclude from their topological analysis that the M···O interactions are 100% ionic in character, implying that the observed magnetic exchange between the metal ions cannot proceed through the super-exchange mechanism. They suggest instead that the strong anti-ferromagnetic coupling involves electron correlation, an idea originally proposed by Slater [61]. If this conclusion were proved to be more general than just for simple metal oxides, then studies on metal–metal bridged species should also be reconsidered.

3 Bond Order $\leq$ 1: Partial Direct Interaction

As indicated in Sect. 1, one of the main problems concerning the electron density analysis of metal–metal bonds is in the relationship between the observed function and a Lewis-type interpretation. Even if some direct exchange is present, an uncertainty remains as to the bond order. In many QTAIM studies, the bond order is estimated through the electron density at the corresponding BCP, which of course relies on the presence of bond paths. As will be clear from the discussions below, the formal bond order is not so straightforwardly linked to a local evaluation of the electron density, and more detailed analyses are necessary. Table 2 lists the experimental and theoretical topological properties for those molecules which have a critical point (not always a BCP) associated with the M–M interaction. In order to retain some comparability, in most cases the theoretical results have been recomputed at the DFT level, using the B3LYP exchange-correlation functional, with optimised geometries. Standard 6-311++G** bases were used for all atoms except the transition metals, for which Wachters+f basis was used.

A number of molecules of general formula $M_m(CO)_n$ with formal *single* M–M bonds (according to the EAN rule) are known. As indicated in Sect. 1, the first ED investigations on such molecules by means of deformation density mapping (either theoretical or experimental) were rather inconclusive. Starting in the late 1980s, however, the QTAIM methodology was applied to such metal complexes, and the

Table 2 Topological properties of M–M formal single bonds[a]

Compound	Refs.	M–M[b]	ρ_{BCP}	$\Delta^2\rho_{BCP}$	λ_1	λ_2	λ_3	ε	G_{BCP}	$(G/\rho)_{BCP}$	V_{BCP}	H_{BCP}	δ(M–M)[c]
$Mn_2(CO)_{10}$[d]	[64]	3.034	0.16	0.06	−0.21	−0.21	0.49	0.0	0.04	0.26	−0.08	−0.04	0.282
	[65]	*2.9042(8)*	*0.190(4)*	*0.815(8)*	*−0.197(3)*	*−0.197(3)*	*1.209(6)*	*0.0*	*0.088*	*0.466*	*−0.12*	*−0.031*	–
		2.9031(2)	*0.144(3)*	*0.720(3)*			*0.81*	*0.1*	*0.07*	*0.45*	*−0.08*	*−0.02*	–
$Co_2(CO)_6(AsR_3)_2$ (R=H)	[70]	*2.6430(2)*	0.271	0.043				0.0	0.093	0.34		−0.09	0.51
(R=Ph)			*0.17*	*1.09*				*0.0*		*0.55*		*−0.02*	–
$Fe_2(CO)_6(\mu\text{–}CO)_3$[e]	[37] [40]	2.532	0.31	2.32	0.10	0.10	2.13	–	0.24	0.79	−0.33	−0.08	0.351
$Co_2(CO)_6(\mu\text{–}CO)_2$[f]	[22]	2.564	0.30	2.73	−0.36	1.43	1.65	–	0.26	0.85	−0.32	−0.07	0.355
$Co_2(CO)_6(\mu\text{–}CO)(\mu\text{–}C_4H_2O_2)$[f, g]	[85]	2.483	0.34	2.58	−0.31	0.75	2.14	–	0.28	0.81	−0.37	−0.10	0.365
	[86]	*2.4402(2)*	*0.46(2)*	*3.4(3)*	*−2.0*	*−1.0*	*6.4*	*1.0*	–	–	–	–	–
	[87]	*2.4222(3)*	*0.76(6)*	*2.0(3)*	*−1.8*	*−1.0*	*4.8*	*0.8*	*0.60*	*0.79*	*−1.06*	*−0.46*	–
$Ru_3(CO)_{12}$[h]	[88]	2.944	0.24	1.03	−0.49	−0.45	1.97	0.08	0.10	0.44	−0.14	−0.03	0.477
		2.852	*0.22*	*2.20*	*−0.46*	*−0.46*	*3.13*	*0.0*	*0.17*	*0.77*	*−0.18*	*−0.01*	–
$Co_3(dpa)_4Cl_2$	[133]	*2.2946(8)*	*0.32(1)*	*5.22(1)*	–	–	–	–	*0.37*	*1.13(3)*	*−0.37*	*0.00*	–
		2.4459(10)	*0.20(1)*	*3.43(1)*					*0.21*	*1.09*	*−0.19*	*0.03*	–

[a]Theoretical studies (if reported) shown on first line in plain font, with experimental results in italic font. Unless otherwise stated, calculations involve DFT optimised geometries using idealised symmetry at the B3LYP/6-311++G**/Wachters+f level. All units in Å, energy densities in Hartree $Å^{-3}$

[b]Metal–metal distance

[c]Delocalisation index between metal atoms $\delta(\Omega_M\text{–}\Omega_M)$

[d]Second row, experimental results from Gervasio et al. [64]; third row experimental results from Farrugia et al. [65]

[e]No M–M bond path, (3,+3) critical point at M–M midpoint

[f]No M–M bond path, (3,+1) critical point at M–M midpoint

[g]Second row, experimental results, triclinic modification [85]; third row, experimental results, orthorhombic modification [86]

[h]Experimental results averaged over three M–M bonds

PhD dissertation of MacDougall [23] contained probably the first topological analyses of M–M bonded molecules. This work was based on minimal basis set HF calculations, without geometry optimisation; results were later reported in [62]. It is interesting to read the comments accompanying those results, for example, that an Mn–Mn bond path was "universally anticipated". This is not mathematically correct, and there is indeed only a serendipitous coincidence between those first very approximate calculations and the more accurate experimental or theoretical studies carried out later. Independent experimental QTAIM studies on $Mn_2(CO)_{10}$ by Bianchi et al. [63, 64] and Farrugia et al. [65] concurred in most details, including the presence of an Mn–Mn bond path, though the two sets of authors differ in their interpretation of the nature of the Mn–Mn bond. The density around the BCP is very flat, leading to the possibility that the molecular graph is not too far from a catastrophe point. An isomeric graph, lacking an Mn–Mn bond path and characterised instead by several (1,3) Mn–C bond paths (see Scheme 2 with the ideal graphs), is an alternative topology, though one which has never actually been observed in practise. Interestingly, this alternative topology would be consistent with the suggestion by Brown et al. [66] that the two $Mn(CO)_5$ moieties are mainly attracted by Mn···CO electrostatic forces, and this thesis was also partially considered when more sophisticated quantum chemical calculations of the dissociation energy were published, see Rosa et al. [67]. However, one shall note that if the source of bonding is a pure electrostatic attraction, then Mn···C or Mn···O bond paths are not a necessity.

In MacDougall's original work, there is not much interpretation on the nature of the Mn–Mn bond, but a comment was made on the strange feature of the Laplacian, which had a slightly negative region at the BCP, because the "two Mn 4s shells merged". With this interpretation, the QTAIM analysis would seemingly address the correctness of a Mn–Mn direct (single) bond. However, later studies by Bianchi et al. [64] and Farrugia et al. [65] demonstrated that the Laplacian is in fact slightly positive at the BCP, as was also evident from more accurate theoretical calculations. Farrugia et al. [65] assigned a weak open-shell, covalent character to the Mn–Mn bond, on the basis of the slightly negative total energy density H_{BCP} and the relatively high integrated density over the Mn–Mn interatomic surface $\oint_{Mn\cap Mn} \rho(\mathbf{r}) = 1.82\ e\text{Å}^{-1}$. Bianchi et al. [64], on the other hand, assigned a closed-shell character to the Mn–Mn bond, although they did not speculate on the consequences of such interpretation and concluded by classifying this as a *metallic* bond, intermediate between ionic and covalent. Their definition of *metallic* bonding [68] clearly raises some issues, as was indeed concluded by Ponec et al. [69] in their

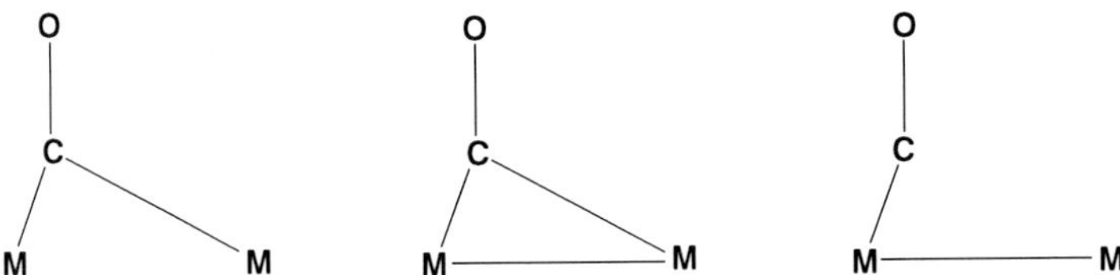

Scheme 2 The possible configurations (and idealised *molecular graphs*) associated with a carbonyl bridging a M–M bond

DAFH study on $Mn_2(CO)_{10}$, who stated that the use of the adjective *metallic* "is unfortunate and misleading because it was introduced and intended for the description of delocalised bonding in bulk metals, which is certainly very different from . . . the bonding in a small cluster . . . containing metal–metal bonds".

Other $M_2(CO)_nL_m$ systems (without ligand bridges) were characterised (experimentally and theoretically), including $Co_2(CO)_6(AsPh_3)_2$ [70], which also showed a Co–Co bond path. In this study, the authors clearly demonstrated that the positive Laplacian could not be the correct indicator to assign a closed- or open-shell bond character, in the presence of transition metals. In particular, the previous work by Cremer and Kraka [71, 72] was important here. They proposed to define a chemical bond as covalent in nature when two conditions were met: (a) a bond path was present and (b) the potential energy density was local overwhelming (implying a negative energy density at the BCP). As with any other criterion resorting to local properties only, this is also questionable.[2] However, it clearly addressed the problem of interpretation of chemical bonding between atoms when the associated BCP has a positive Laplacian. Macchi et al. [70] first applied the energy density criterion to M–M bonds in their experimental charge density study on $Co_2(CO)_6(AsPh_3)_2$, noting in fact the fulfilment of the Cremer and Kraka conditions. Despite the compound and its stereochemistry being quite different, the Co–Co bond showed, not surprisingly, many similarities with the Mn–Mn bond in $Mn_2(CO)_{10}$.

An interesting test of the electron density and energy density criteria on M–M bonds comes from the comparative study by Novozhilova et al. [73] on the anion $[Pt_2(H_2P_2O_5)_4\,]^{4-}$ in the ground and excited triplet states. The ground state is formally missing an M–M bond, whereas in the first excited state an M–M bonding orbital would be partially populated, giving a formal bond order of 0.5. Experiments carried out under laser excitation do in fact demonstrate considerable shortening of Pt–Pt bond distances [74], in agreement with theoretical predictions that also address an increasing Pt–Pt bonding density and a more negative energy density. However, it is notable that the ground state also shows a Pt–Pt bond path associated with a small but negative energy density (Table 5).

In an in-depth review, Macchi and Sironi [75] suggested that the electron density shared between two metals be used in addition to analyse M–M bonds. They proposed two main response indicators: (a) the electron density integrated over the interatomic surface and (b) the electron delocalisation between the two atomic basins. The first quantity has the advantage that it only depends on the electron density (so in principle it can be evaluated from an experimental multipolar density). This idea is somewhat related to Berlin's theorem [76, 77]. The delocalisation index is more inherently linked to the concept of electron sharing, hence covalency. The interesting findings arising from these studies were that apparent single M–M bonds (even if not supported by ligand bridges) have only a fractional bond order (usually 0.3–0.5; see Table 2). In addition, all M···C interactions, even if

[2] If the electrons were a classical fluid, then the energy density would be the pressure exerted on the electrons. Regions of negative pressure attract electrons.

not associated with a bond path in the molecular graph, have some significant electron sharing. All these conclusions were, not surprisingly, confirmed by analysis of the DAFH. Indeed, as the DAFH is grounded in the pair density (as are the delocalisation indices), Ponec et al. [69] could clearly visualise the partially direct (1,3) Mn···CO interactions in $Mn_2(CO)_{10}$. In a way, the DAFH could be considered as a pictorial representation of the delocalisation indices.

The new picture drawn from what we could call "deeper" QTAIM analysis provides a different interpretation of the electron density, with respect to a more "classical" analysis. We can summarise the main points as follows:

(a) M–M bonds are never entirely localised, unless a pure gas phase metal dimer is considered. The degree of delocalisation rises if the ligands around the metals are Lewis acidic.
(b) In metal carbonyls, (1,3) M···C interactions contribute to the M–M linkage not only because of favourable electrostatic interaction, but also thanks to genuine electron sharing.
(c) The electron counting rules, based on the assumption of localised M–M bonds (at least in lower nuclearity cluster), are clearly an over-simplification.

The picture is even clearer if bridged systems are studied (Tables 1 and 2). Experimental studies on these systems include the carbonyl-bridged systems $Co_4(CO)_{11}(PPh_3)$ [78], $[FeCo(CO)_8]^-$ [79], the alkylidyne-bridged complexes $Co_3(\mu_3\text{–}CX)(CO)_9$ (X=H, Cl) [80], the borylene-bridged complex $Mn_2(\eta^5\text{–}C_5H_5)_2(CO)_4(\mu\text{–}B^tBu)$ [81, 82], and alkyne-bridged species $Co_2(CO)_6(\mu\text{–}\eta^2\text{–}HC{\equiv}CC_6H_{10}OH)$ [83] and $Co_4(CO)_{12}(\mu_4\text{–}\eta^4\text{–}PhC{\equiv}C\text{–}C{\equiv}CPh)$ [84]. Although in all these cases, single M–M bonds are required by the EAN rule, and the molecular graphs consistently lack any M–M bond paths (and usually *any* type of critical point associated with the M–M interaction). Moreover, a large delocalisation is found between these "non-bonded" metal atoms, substantially similar to that in unsupported systems. One apparent exception is the complex $Co_2(CO)_6(\mu\text{–}CO)(\mu\text{–}C_4H_2O_2)$, which was originally reported by Gervasio and co-workers to possess a Co–Co bond path in both its triclinic [85] and orthorhombic [86] modifications, from experimental multipole refinements. It has a bridging CO ligand and a 5-oxofuran-2(5*H*)-ylidene ligand and a formal single Co–Co bond required by the EAN rule, like $Co_2(CO)_8$. However, a subsequent study [87] strongly indicated that this molecule has no Co–Co bond path, but instead possessed a ring structure. The reason for this discrepancy is probably because the system is close to a catastrophe point, where any slight modification of the density model can result in a topological change (see [83, 84] for an extensive discussion of this regarding Co–C(alkyne) interactions).

The current state of affairs is therefore that all reported organometallic species, with a supported, formal single M–M bond, and *at their equilibrium geometry*, do not show a BCP between the metal centres. However, some caveats ought to be issued here. First of all, recent work by Ponec and Gatti [42] has shown that the topology of the electron density in such molecules may be strongly affected by an artificial shortening of the equilibrium M–M distance, leading to the appearance of M–M bond paths. Second, as demonstrated below, when the formal multiplicity of the M–M bond is greater than one, then an M–M bond path is normally observed. Continuous QTAIM indicators like the delocalisation index and the DAFH vary smoothly over the topological catastrophe and seem to be preferable measures of the chemical bonding in metal–metal bonded systems.

As shown by Macchi and Sironi [75], upon transformation from unbridged to bridged, the electron delocalisation through a bridging carbonyl is substantially altered. For this reason, these authors spoke of "interplay between direct and indirect M–M and M–C bonding", in which singular contribution is of course geometry dependent. The analysis of the Laplacian of the carbonyl is particularly interesting, as one can clearly see the polarisation of the lone pair density from a single, localised OC→M donation into a delocalised donation. In fact the associated VSCC is typically re-directed towards the middle of an M–M bond, in symmetric bridge geometries. For the carbonyl ligand then, a single VSCC is associated with the ligand bonding, either to a single metal atom or to more than one (Fig. 2b).

On the other hand, for the alkylidyne carbon atom in the complexes $Co_3(\mu_3\text{–}CX)(CO)_9$ (X=H, Cl) [80], there are four VSCCs, which is an indicative of a localised bonding to the three Co atoms (Fig. 3).

In the case of the borylene-bridged complexes $Mn_2(\eta^5\text{–}C_5H_5)_2(CO)_4(\mu\text{–}BR)$, the evidence regarding the VSCCs on the B atom is ambiguous. In the experimental study [81], two VSCCs were observed for the two Mn–B interactions (Fig. 2a), but in the theoretical study [82], generally only one VSCC is associated with the interaction, though there is considerable sensitivity to the level of computation. It appears that catastrophe situations in the Laplacian of ρ can also arise (as is often seen for the density itself). This in turn leads to an ambiguity in the designation of the ligand, as a borylene (with one Mn···B VSCC) or as a substituted borane (with one VSCC per Mn···B interaction).

A high degree of fluxionality is typically observed in $M_m(CO)_n$ species, implying a relatively flat energy landscape. This results in, for instance, a facile transformation

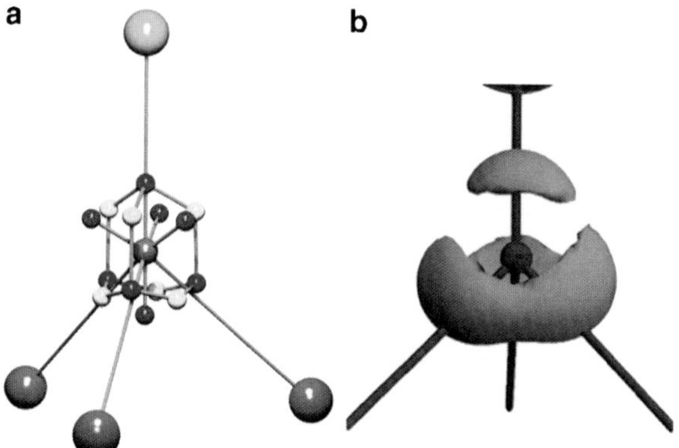

Fig. 2 Plots of the Laplacian $L \equiv -\Delta^2(\rho)$ in the $M_2(\mu\text{–}L)$ plane for (**a**) $Mn_2(\eta^5\text{–}C_5H_5)_2(CO)_4(\mu\text{–}BR)$ R = ^{t}Bu (experimental) and (**b**) $Fe_2(CO)_9$ (theoretical). Positive contours as *solid lines* and negative contours as *broken lines*

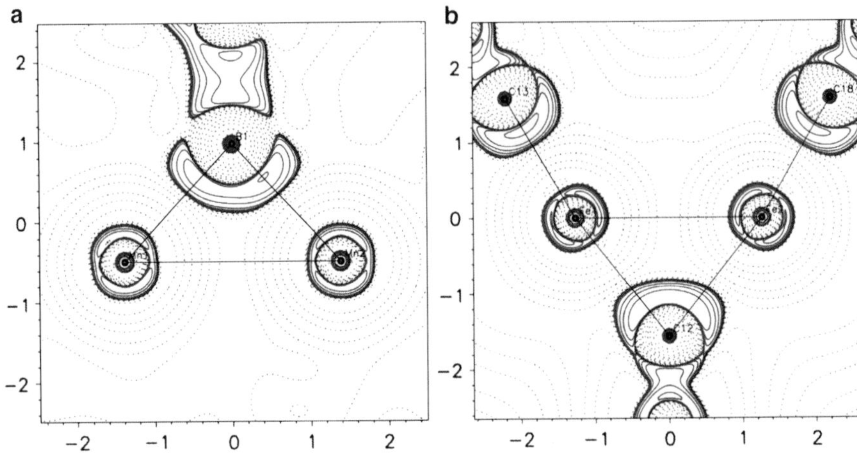

Fig. 3 Laplacian of the electron density for the alkylidyne carbon atom in $Co_3(\mu_3\text{–}CCl)(CO)_9$. (**a**) Atomic graph, i.e. critical points in the valence shell charge concentrations and (**b**) isosurface (+10 eÅ^{-5}) of $-\nabla^2\rho(\mathbf{r})$

between terminal and bridging carbonyls. It is noteworthy that when interpreted in terms of the electron density, fluxionality implies only one kind of discontinuity, namely that in the molecular graph. All other indicators [δ(M–M), DAFHs, electron density, Laplacian and VSCCs] are continuously transformed, without abrupt changes, which seems more reasonable when considering the potential energy surface associated with these molecules. From another point of view, the observed localised alkylidyne–cobalt bonding in the topology of $Co_3(\mu_3\text{–}CX)(CO)_9$ (X=H, Cl) [80] can be used to provide a clear rationalisation of the facile fluxional tripodal rotation of the $Co(CO)_3$ groups, since a low barrier to rotation about the Co–C single bonds is expected.

Presumably, at least partly, due to experimental difficulties, there have been very few QTAIM studies on metal cluster compounds with more than four metal atoms. As far as we are aware, only one experimental study has been undertaken on metal–metal bonding outside the first transition series in the compound $Ru_3(CO)_{12}$ [88]. Three Ru–Ru bond paths were observed (as expected for this unbridged system), but also unusual C···C bond paths between the adjacent axial carbonyl groups, which was taken as justification for the non-linearity of the axial Ru–CO bonds. These C···C interactions are not observed in theoretical studies; so their interpretation remains unclear at present. Macchi et al. [78] analysed the electron density of $Co_4(CO)_{11}PPh_3$, a tetrahedral cobalt cluster characterised by three carbonyl-bridged and three unsupported Co–Co edges, all formal single bonds according to the EAN rules. The molecular graph is in agreement with the absence of bond paths for the bridged Co–Co bonds; thus, only three M–M paths are located, from theory and from experiment. All the QTAIM features of the Co–Co bonds are similar to the case of bimetallic compounds. Further preliminary work by Macchi and Sironi [89] includes some octahedral Co_6 clusters, some of them containing interstitial hydrides or carbides. Despite the completely different nature of these clusters, the general "topological rule" that bond paths disappear when formally single bonds are bridged is also fulfilled – an interstitial atom in a octahedral cluster being basically a bridge for all of them (Scheme 3). Any interpretation of delocalisation indices in these systems is clearly very complicated.

Finally, there have been two very interesting experimental studies on unbridged M–M bonded systems, which differ fundamentally from those described above, in which no *d*-electrons are formally involved in the bonding. These are the main group complexes $Zn_2(\eta^5\text{–}C_5Me_5)_2$ [90] and the β-diketiminate complex $Mg_2(\text{dippnacnac})_2$ [91] (dippnacnac = $(ArNCMe)_2CH$, Ar =2,6-$^iPr_2C_6H_3$), which both involve the M_2^{2+} cation in the unusually low M(I) oxidation state. Both studies show similar topological characteristics, with very low values of ρ_{BCP} and $\Delta^2\rho_{BCP}$ and a small, slightly negative value for H_{BCP}. These properties may be attributed to the highly diffuse nature of the M–M bonds, being σ-bonds of primarily *s*-type character [92–94]. In the case of the Zn_2 compound, the integrated density over the Zn–Zn interatomic surface $\oint_{Zn\cap Zn}\rho(\mathbf{r})$is 1.25 $e\text{Å}^{-1}$, which shows that the bond is not necessarily a particularly weak one [90]. This is corroborated by the delocalisation index $\delta(\Omega_{Zn}\text{–}\Omega_{Zn})$ between the two metals, which is very close to 1.0. In $Mg_2(\text{dippnacnac})_2$, on the other hand, the bond appears much weaker, despite an energy density profile along the bond path clearly revealing a large area where the potential energy density overwhelms. The Mg–Mg distance is quite long, probably because of large steric hindrance between the two ligands, and this might affect the Mg–Mg bonding.

We have calculated the Mg_2^{2+} and Zn_2^{2+} dications in isolation and coordinated to the same μ-diketiminate ligand $[(HNCMe)_2CH)]^-$, with the aim of comparing the M(I)–M(I) bonding. These results are summarised in Table 3 and Fig. 4.

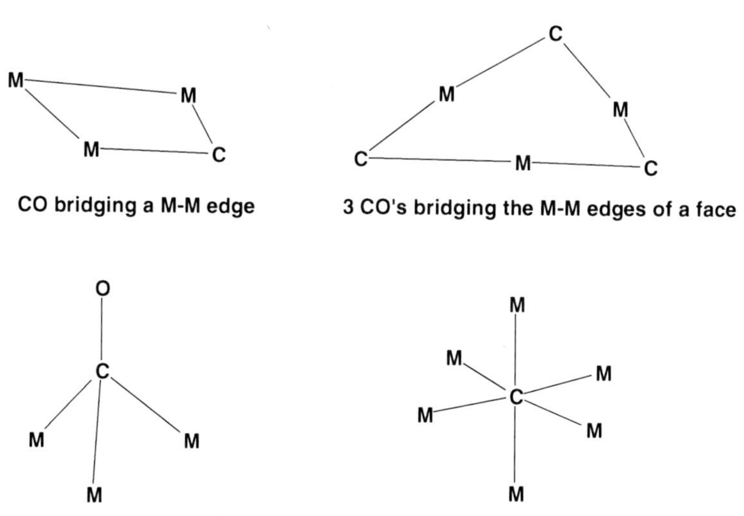

Scheme 3 Idealised *molecular graphs* in some higher nuclearity cluster

Table 3 Topological properties of formal M(I)–M(I) metal–metal single bonds[a]

Compound	Refs.	M–M[b]	ρ_{BCP}	$\nabla^2\rho_{BCP}$	λ_1	λ_2	λ_3	ε	G_{BCP}	$(G/\rho)_{BCP}$	V_{BCP}	H_{BCP}	δ(M–M)[c]
$Zn_2(\eta^5\text{–}C_5Me_5)_2$[d]	[90]	2.1657	0.426	1.622	–	–	–	–	0.267	0.627	−0.42	−0.154	0.919
		2.3186(3)	*0.348(3)*	*1.824(17)*	–	–	–	–	–	–	–	–	–
$Zn_2[(HNCMe)_2CH)]_2$	–	2.398	0.406	0.619	−1.03	−1.03	2.69	0.0	0.198	0.48	−0.353	−0.155	0.918
Zn_2^{2+}	–	2.5246	0.225	1.471	−0.79	−0.79	3.05	0.0	0.136	0.607	−0.170	−0.033	1.02
$Mg_2(dippnacnac)_2$[e]	[91]	*2.8456(2)*	*0.10(2)*	*0.261(3)*	*−0.32*	*−0.26*	*0.84*	*0.23*	*0.01*	*0.30*	–	*−0.01*	–
$Mg_2[(HNCMe)_2CH)]_2$[f]	–	2.873	0.196	−0.722	–	–	–	–	0.005	0.024	−0.06	−0.055	0.279
Mg_2^{2+}	–	2.926	0.172	−0.573	−0.311	−0.311	0.05	0.0	0.003	0.0167	−0.046	−0.043	1.0

[a]Theoretical studies (if reported) shown on first line in plain font, with experimental results in italic font. Unless otherwise stated, theoretical calculations are previously unpublished involving DFT optimised geometries, at B3LYP/6-311++G(2d,2p) level, All units in Å, energy densities in Hartree $Å^{-3}$

[b]Metal–metal distance (experimental values in italics)

[c]Delocalisation index between metal atoms $\delta(\Omega_M\text{–}\Omega_M)$

[d]Calculation at MP2/6311G(d,p) level

[e]Experimental results only, no theoretical data available

[f]Computed at the non-nuclear attractor (midpoint of Mg–Mg bond path)

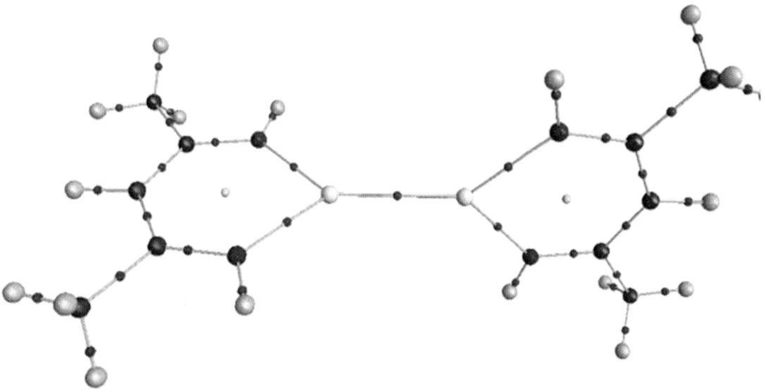

Fig. 4 Theoretical molecular graph of $Zn_2[(HNCMe)_2CH)]_2$

In $M_2[(HNCMe)_2–CH)]_2$, Mg and Zn behave as in $Mg_2(dippnacnac)_2$ and $Zn_2(\eta^5–C_5Me_5)_2$: the Mg–Mg distance is similar to the isolated dication, whereas the Zn–Zn bond distance is shorter, which induces a larger electron density and a larger electron delocalisation. Notably, in $Mg_2[(HNCMe)_2CH)]_2$, computed at B3LYP/6-311G(2d,2p) level, an NNA is found at the Mg–Mg midpoint, lying in a region of flat electron density. The experimental study by Overgaard et al. [91] for $Mg_2(dippnacnac)_2$ confirms a very flat region of electron density, though a Mg–Mg BCP is found. If compared with the isolated Mg_2^{2+}, we note a large difference in the Mg–Mg delocalisation, which is close to 1.0 in Mg_2^{2+}, but much smaller in $Mg_2[(HNCMe)_2CH)]_2$. Thus, it seems that coordination weakens the Mg–Mg interaction (at least it reduces the Mg–Mg delocalisation), whereas it does not affect the Zn–Zn interaction, in which distance is even shortened. Notably, Datta [95] reported calculations on the hydrogenation of Mg(I)–Mg(I) complexes, which should be thermodynamically favourable, especially because of the more stable delocalised $Mg(\mu–H)_2Mg$ system.

4 Bond Order $\geq$ 1: Multiple M–M Bond with and Without Bridges

In this section, we tackle the thorny problem of compounds with metal–metal bond orders greater than unity. The reason for this "thorniness" is the great difficulty in precisely defining the term "bond order", particularly in relation to M–M multiple bonds. As an example, we need look no further than the classic textbook [96] case of the $[Re_2Cl_8]^{2-}$ anion, which is considered to have a formal quadruple bond, on the basis of the $\sigma^2\pi^4\delta^2$ di-rhenium *d*-electron configuration. However, more recent sophisticated quantum mechanical treatments of this anion using multiconfigurational wavefunctions including relativistic corrections and spin–orbit coupling [97, 98], energy decomposition analysis [99] or analysis of the DAFH [100] conclude that the effective bond order is closer to three. Moreover, for the formally quadruple W–W bond in $[W_2Cl_8]^{4-}$, Macchi and Sironi [75] computed a delocalisation index $\delta(\Omega_W{-}\Omega_W)$ close to 3.0.[3] A similar ambiguous situation pertains to the formal triple bond in the $[RGa{\equiv}GaR]^{2-}$ dianion, as is summarised by Bader [101]. The delocalisation index $\delta(\Omega_{Ga}{-}\Omega_{Ga})$ is indicative of a bond order close to two rather than three, and the lack of correlation between bond lengths, bond strengths, bond orders and QTAIM topological indicators such as ρ_{BCP} and $\Delta^2\rho_{BCP}$ when dealing with M–M bonds was also emphasised in this work [101], as it was earlier by Macchi and Sironi [75]. Ponec et al. [102] came to a similar conclusion regarding the multiplicity of the Ga–Ga bond from DAFH studies.

We initially compare the theoretical topological properties for a small set of diamagnetic Cr_2 complexes – this metal being chosen because it displays formal Cr–Cr bond orders ranging from 1 to 5 in such complexes. The results are summarised in Table 4, listed in order of increasing $\delta(\Omega_{Cr}{-}\Omega_{Cr})$. The most obvious conclusion is that a Cr–Cr bond path is almost always observed, even when the Cr–Cr bond is bridged by multiple ligands, in marked contrast to the situation described in the previous section. The only exception in Table 4 is for $Cr_2Cp_2(\mu{-}S)_2(\mu{-}\eta^2{-}S_2)$, which has a ring critical point close to the centre of the rather long Cr–Cr vector (2.764 *theor*, 2.807(1)Å *expt*, [103]). This conclusion even applies to $Cr_2Cp_2(CO)_6$, which has an unusually long Cr–Cr distance and a very weak Cr–Cr single bond [104]. However, in this case, in addition to the Cr–Cr bond path, with a very low ρ_{BCP} value of 0.12 e, there are other bond paths linking the two $CrCp(CO)_3$ fragments, arising from four CH···O and two C···C interactions (Fig. 5).

The value of $\delta(\Omega_{Cr}{-}\Omega_{Cr}) = 0.272$ indicates that much less than one electron pair is shared between the Cr atoms, but analysis of the orbitals contributing to this index indicates that some 96% of the electron pair exchange occurs through a single orbital (shown in Fig. 6), which is clearly associated with a direct Cr–Cr σ-bond. The other orbitals that contribute to $\delta(\Omega_{Cr}{-}\Omega_{Cr})$ are delocalised over the carbonyl ligands, in line with the suggestions of Macchi and Sironi [75, 79] outlined above. The QTAIM analysis therefore indicates that in $Cr_2Cp_2(CO)_6$, the two 17e fragments are held together not only by some weak inter-fragment interactions but also by a (presumably very weak) direct Cr–Cr bond.

A second conclusion from the data in Table 4 is that, of all the topological indicators, the delocalisation index provides the clearest link to the bond order, though it is obviously not related in any way directly. We will return to this issue later. First, we will discuss a recent "hot topic" in multiple M–M bonds, i.e. the formal quintuple bond. A number of compounds containing very short, formal quintuple bonds between chromium(I) centres have been reported, including the bis-arene complexes [ArCrCrAr] of Power and co-workers [105] (Cr–Cr = 1.8351 (4) Å), the bis(diazadiene) $Cr\{\mu{-}\eta^2{-}NR(H)C{=}C(H)NR\}_2Cr$ of Kreisel et al. [106]

[3] In the original publication by Macchi and Sironi [72], the compound formula is erroneously reported as a dianion, whereas calculations were in fact carried out on the quadruple-bonded tetra-anion.

Table 4 Topological properties of metal–metal bonds in dimeric chromium complexes[a]

Compound	FBO[b]	M–M[c]	ρ_{BCP}	$\nabla^2\rho_{BCP}$	λ_1	λ_2	λ_3	ε	G_{BCP}	$(G/\rho)_{BCP}$	V_{BCP}	H_{BCP}	δ(M–M)[d]
$Cr_2Cp_2(CO)_6$	1	3.376 *3.26175(9)*	0.12	0.14	−0.17	−0.05	0.36	2.44	0.03	0.21	−0.04	−0.02	0.272
$Cr_2Cp_2(CO)_2(\mu\text{–}PMe_2)_2$	2	2.639 *2.5776(7)*	0.32	0.94	−0.80	−0.20	1.93	3.11	0.17	0.52	−0.27	−0.10	0.569
$[Cr_2Cp_2(\mu\text{–}\eta^5\text{–}P_5)]^-$	?	2.616	0.38	0.58	−0.64	−0.64	1.86	0.00	0.17	0.45	−0.31	−0.13	0.573
$Cr_2(\eta^6\text{–}C_6H_6)_2(\mu\text{–}CO)_3$	3	2.240 *2.221(av)*	0.60	5.34	−2.64	−2.64	10.63	0.00	0.55	0.92	−0.73	−0.18	0.593
$Cr_2Cp_2(\mu\text{–}S)_2(\mu\text{–}\eta^2\text{–}S_2)$[e]	?	2.764 *2.807(1)*	0.30	1.02	−0.52	0.69	0.86	0.00	0.15	0.50	−0.23	−0.08	0.677
$Cr_2Cp_2(CO)_4$	3	2.236 *2.215(av)*	0.58	4.61	−2.22	−2.20	9.02	0.01	0.51	0.87	−0.70	−0.19	1.022
$Cr_2Cp_2(\mu\text{–}\eta^8\text{–}C_8H_8)$	3	2.353 *2.390(2)*	0.51	2.39	−1.55	−1.49	5.43	0.04	0.33	0.65	−0.50	−0.16	1.174
$Cr_2Cp_2(\mu\text{–}\eta^3\text{–}C_3H_5)_2$	4	2.297 *2.299(1)*	0.57	3.27	−1.93	−1.62	6.82	0.19	0.41	0.73	−0.60	−0.18	2.045
$Cr_2(\eta^4\text{–}C_8H_8)_2(\mu\text{–}\eta^8\text{–}C_8H_8)$	4	2.044 *2.213(x)*	0.91	7.97	−4.47	−3.95	16.39	0.13	0.95	1.04	−1.33	−0.39	2.055
$Cr_2(\eta^3\text{–}C_3H_5)_2(\mu\text{–}\eta^3\text{–}C_3H_5)_2$	4	1.869 *1.971(x)*	1.28	16.03	−7.52	−7.14	30.69	0.05	1.89	1.47	−2.66	−0.77	2.680
$Cr_2(O_2CR)_4$ (R=H)[f] (R=2,4,6–$^iPr_3C_6H_2$)	4	1.966 *1.9662(5)*	1.06	12.42	−4.66	−4.66	21.74	0.0	1.39	1.31	−1.90	−0.52	3.215
$Cr_2\{RNC(H)=C(H)NR\}_2$ (R=H)[g] (R=1,2–$^iPr_2Ph$)	5	1.764 *1.8028(9)*	1.65	24.04	−9.97	−9.96	43.96	0.0	2.92	1.77	−4.15	−1.24	3.751

[a]Theoretical studies (this work). Unless otherwise stated, these involve DFT optimised geometries using idealised symmetry at the B3LYP/6-311++G**/Wachters+f level. All units in Å, energy densities in Hartree $Å^{-3}$

[b]Formal bond order (normally based on EAN rule)

[c]Metal–metal distance (experimental values in italics)

[d]Delocalisation index between metal atoms $\delta(\Omega_M\text{–}\Omega_M)$

[e]No M–M bond path, (3,+1) critical point at Cr–Cr midpoint

[f]Single point calculation using D_{4h} symmetrised geometry from $Cr_2(O_2CAr)_4$ Ar=2,4,6–$^iPr_3C_4H_2f$

[g]Single point calculation using coordinates from [106]

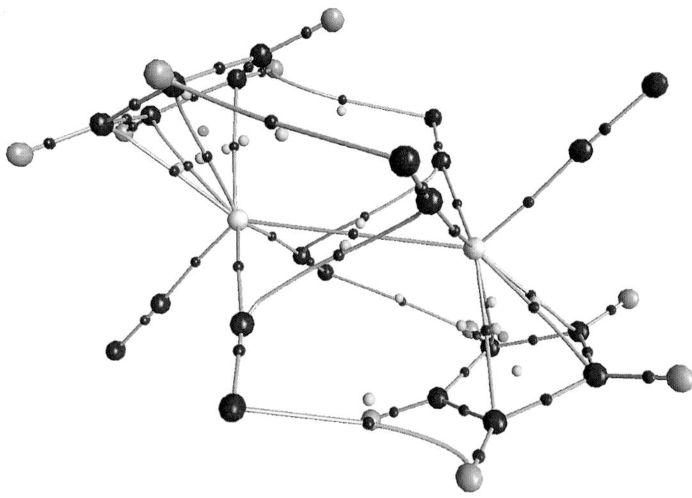

Fig. 5 Theoretical molecular graph of $Cr_2Cp_2(CO)_6$

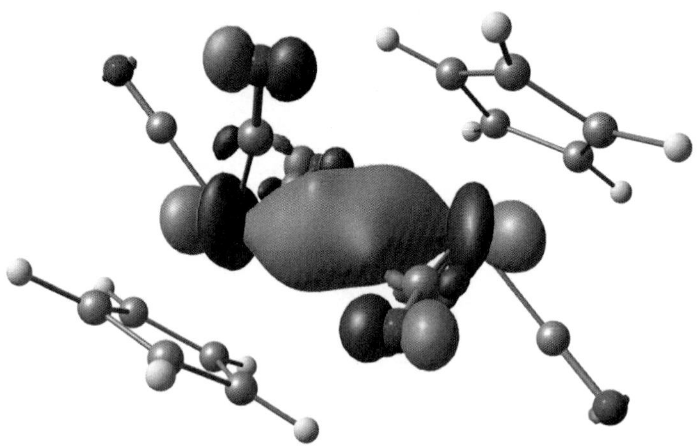

Fig. 6 Canonical Kohn–Sham orbital contributing 96% towards the delocalisation index $\delta(\Omega_{Cr}–\Omega_{Cr})$ in $Cr_2Cp_2(CO)_6$

(Cr–Cr = 1.8028(9) Å), the tris(amidate) anion $[Cr\{\mu–\eta^2–NR–CH–NR\}_3Cr]^-$ (Cr–Cr = 1.8169(7) Å) and the bis(amidate) $Cr\{\mu–\eta^2–NR–CR'–NR\}_2Cr$ (Cr–Cr = 1.7404(8) Å) of Tsai and co-workers [107, 108] and the aminopyridinato complex $Cr\{\mu–\eta^2–NR–pryridyl\}_2Cr$ (Cr–Cr = 1.749(2) Å) of Noor et al. [109]. Recently, Tsai and co-workers [110] have reported an analogous bis(amidate) molybdenum complex $Mo\{\mu–\eta^2–NR–CR'–NR\}_2Mo$ with a very short, formal quintuple Mo–Mo bond (Mo–Mo = 2.0875(4) Å).

The synthesis of these complexes has resulted in an explosion of theoretical papers and mini-reviews on the nature of formal quintuple (and higher multiplicity)

M–M bonds [106–122]. The majority of the theoretical studies have focussed on orbital-based interpretations of the wavefunction, and the earlier studies in particular used relatively simple DFT approaches to verify the quintuple nature of the bond, based on a $\sigma^2\pi^4\delta^4$ electron configuration. For instance, Kreisel et al. [106] estimated an effective bond order of 4.28 from an NBO approach. However, more sophisticated calculations using multiconfigurational wavefunctions including relativistic and spin–orbit coupling corrections [113, 116, 118, 122] invariably result in smaller effective bond orders, in the range 3.3–3.8, due to the poor involvement of the δ orbitals in the M–M bond.

There have been a few studies of quintuple M–M bonding involving real-space analysis of the electron density. Noor et al. [109] have studied simplified models of their aminopyridinato compound using a DFT wavefunction and the pELI-D approach [123]. The σ-MO, the two π-MOs and one of the δ-MOs have their pELI-D maxima between the Cr atoms, while the second δ-MO has four pELI-D maxima close to the Cr atoms. This is consistent with one weakly bonding δ-orbital, as is the delocalisation index $\delta(\Omega_{Cr}–\Omega_{Cr})$ of 4.2. Dupré [117] has undertaken a QTAIM analysis of the DFT density for a model of the bis(diazadiene) complex of Kreisel et al. [106]. The topological parameters are given in Table 4 and show somewhat atypical values, especially for the energy densities G_{BCP} and V_{BCP} due to the very short Cr–Cr separation of 1.764 Å. The delocalisation index $\delta(\Omega_{Cr}–\Omega_{Cr})$ of 3.6 is somewhat smaller that reported by Noor et al. [109], but again indicates an actual bond order that is considerably smaller than the formal bond order. Ponec and Feixas [120] have also studied the bis(diazadiene) complex by analysis of the DAFH and come to a similar conclusion. Four eigenvalues of the DAFH can be assigned to electron pairs in localised Cr–Cr bonds, but the fifth eigenvalue is much less than 2.0 and is strongly delocalised over the ligand, resulting in a much weakened Cr–Cr bond. The view of these authors is that there is effectively only a quadruple bond present.

Gatti and Lasi [124] have analysed a series of M–M bonds, ranging from formal zero order to highly unsaturated M–M bonds. They analysed the electron density indicators, in particular testing the information available from the source function (SF). The SF, originally defined by Gatti and Bader [125], allows a reconstruction of the density at some (arbitary) reference point. The SF relies only on the knowledge of the density and its derivatives and so is available from both experimental and theoretical densities (unlike the delocalisation index, for example). For this reason, a number of studies have investigated the chemical information present in the SF [126, 127]. In the series $Co_2(CO)_x$ ($x = 8,7,6,5$), Gatti and Lasi [124] tested Co–Co bonds of formal order 1–4 respectively. It is interesting that triply and quadruply Co–Co bonded molecules, though bearing a CO bridge, are anyway characterised by a metal–metal bond path, as we have seen above. On the other hand, the formal BO does not correlate well with topological indicators, and even the formally quadruple bond results in a quite small delocalisation index $\delta(\Omega_{Co}–\Omega_{Co})$, less than 1.0. Nevertheless, delocalisation increases with the formal BO and, by analogy, the relative SF contributions from the two metal increase at the BCP (or at the Co–Co midpoint, for molecular graphs lacking of a Co–Co bond path).

Table 5 Topological properties of bimetallic complexes of second and third row transition metals[a]

Compound	Refs.	FBO[b]	M–M[c]	ρ_{BCP}	$\nabla^2\rho_{BCP}$	λ_1	λ_2	λ_3	ε	G_{BCP}	$(G/\rho)_{BCP}$	V_{BCP}	H_{BCP}	δ(M–M)[d]
$Nb_2(HNCHNH)_4$	[128]	3	2.224	0.147	0.392	–	–	–	–	0.178	1.21	−0.257	−0.079	1.025
$Mo_2(HNCHNH)_4$	[128]	4	2.092	0.185	0.550	–	–	–	–	0.25	1.35	−0.363	−0.113	1.255
$Tc_2(HNCHNH)_4$	[128]	3	2.082	0.179	0.609	–	–	–	–	0.245	1.37	−0.337	−0.092	1.371
$Ru_2(HNCHNH)_4$	[128]	2	2.493	0.079	0.168	–	–	–	–	0.054	0.68	−0.067	−0.013	0.551
$Rh_2(HNCHNH)_4$	[128]	1	2.459	0.072	0.158	–	–	–	–	0.053	0.74	−0.066	−0.013	0.373
$Pd_2(HNCHNH)_4$	[128]	0	2.691	0.042	0.115	–	–	–	–	0.039	0.93	−0.050	−0.011	0.124
$[Pt_2(H_2P_2O_5)_4]^{4-}$ ground state[e]	[73]	0	3.039	0.236	0.79	−0.58	−0.58	1.93	0.0	0.081	0.343	−0.107	−0.027	–
$[Pt_2(H_2P_2O_5)_4]^{4-}$ excited state[e]	[73]	0.5	2.839	0.344	2.94	−0.89	−0.89	2.94	0.0	0.148	0.431	–	−0.067	–
$Rh_2(CO)_2(H_2PCH_2PH_2)_2$[f]	[134]	1	2.878	0.28	0.60	–	–	–	–	–	–	–	–	–

[a]Theoretical studies involve DFT optimised geometries using idealised symmetry at the B3LYP/3-21G** level, unless otherwise stated. All units in Å, energy densities in Hartree $Å^{-3}$

[b]Formal bond order (based on EAN rule)

[c]Metal–metal distance

[d]Delocalisation index between metal atoms $\delta(\Omega_M–\Omega_M)$

[e]For calculation details, see [73]

[f]For calculation details, see [134]

Llusar et al. [128] and later Gatti and Lasi [124] analysed the D_{4h} complexes $M_2(HNCHNH)_4$ (M being a second row transition metal, from Nb to Pd). The results are listed in Table 5. This series formally spans formal bond orders from 0 (Pd) to 4 (Mo), while the stereochemistry of the molecules remaining relatively unchanged (Fig. 7). Llusar et al. [128] mainly analysed the system using the electron localisation function (ELF) and found a surprisingly large covariance between the metal cores, interpreted in terms of resonance. Gatti and Lasi [124] instead used QTAIM and, in particular, delocalisation indices and the source function. At variance from the ELF, QTAIM does not distinguish core and valence domains; however, a large electron delocalisation was computed, and Gatti and Lasi also noted the correlation with SF contribution from the metal atoms at the BCP. In particular, Gatti and Lasi stressed that many "classical" QTAIM indices, all based on the evaluation of local properties, fail when trying to explain the M–M multiple bonds.

It is quite apparent from the data in Tables 4 and 5 that the classic topological parameters are not of much utility in defining a metal–metal bond order. The most useful and most direct indicator is the delocalisation index $\delta(\Omega_M\text{–}\Omega_M)$, but even this has often little connection to the formal bond order. This is because this index carries information about both the *strength* and the *multiplicity* of the bond. Although these two terms are obviously connected, they are not synonymous, and

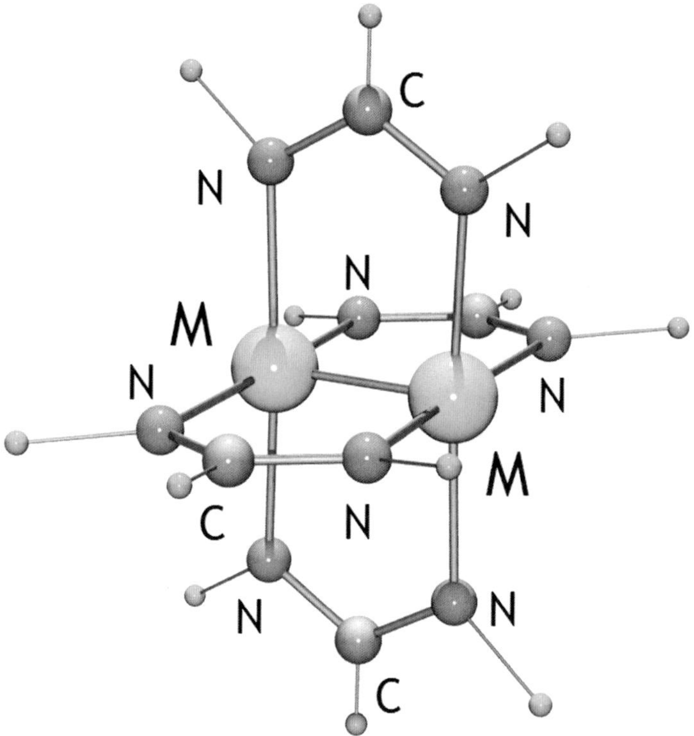

Fig. 7 Structure of the formidamate complexes $M_2(HNCHNH)_4$ (M = Nb, Mo, Tc, Ru, Pd)

we can envisage the situation where a strong single bond has similar topological indicators to a weak double bond. Some further insight into the significance of the delocalisation index may be afforded by examination of its orbital contributions. For example, we can deconvolute the delocalisation index $\delta(\Omega_{Cr}–\Omega_{Cr}) = 3.751$ for the formally quintuply bonded complex $Cr_2\{HNC(H)=C(H)NH\}_2$ into contributions from the canonical Kohn–Sham orbitals, shown in Fig. 8. MOs 51 (σ-Cr–Cr bond), 49 and 50 (π-Cr–Cr bond), and 53 (δ-Cr–Cr bond) are all highly localised on the two

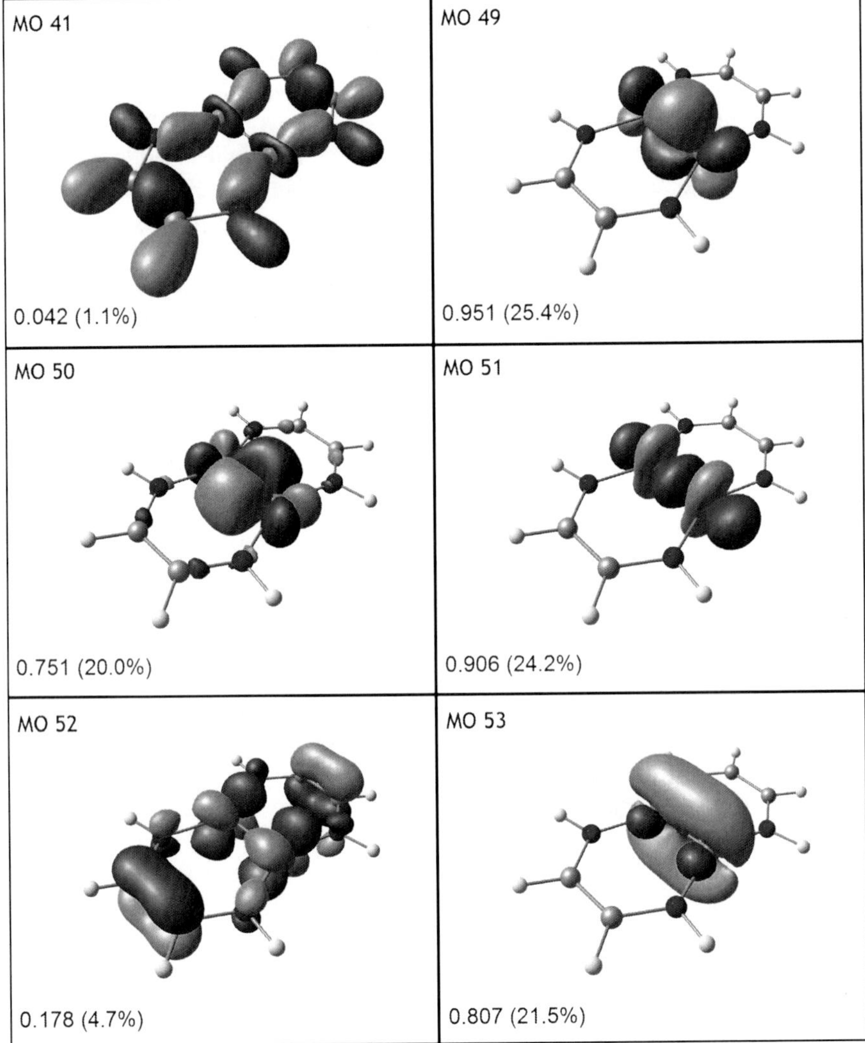

Fig. 8 The six Kohn–Sham canonical orbitals of $Cr_2\{HNC(H)=C(H)NH\}_2$ providing the major contributions to the delocalisation index $\delta(\Omega_{Cr}–\Omega_{Cr})$. The absolute and percentage contributions to the delocalisation index are indicated

Cr atoms. MO 52 partly provides another δ-Cr–Cr bond, but there is substantial delocalisation onto the ligand atoms, as is also the case for MO 41. The true Cr–Cr bond order is therefore significantly less than five, according to this qualitative analysis, which happily agrees with the more quantitative analyses on the BO, discussed above for this type of M–M bond.

Finally, we note that it is apparent that the polarisation in the VSCC of the metal atoms is dominated by the ancillary ligands, rather than by the M–M interactions. In fact, an examination of the complexes in Table 4 quickly shows that there is no clear-cut polarisation of the metal atom associated with even quite strong M–M bonds. Figure 9 shows plots of the Laplacian $L \equiv -\nabla^2(\rho)$ in the plane of the Cr–Cr bond for four representative complexes. In some cases, as for $Cr_2(O_2CH)_4$ (Fig. 9a)

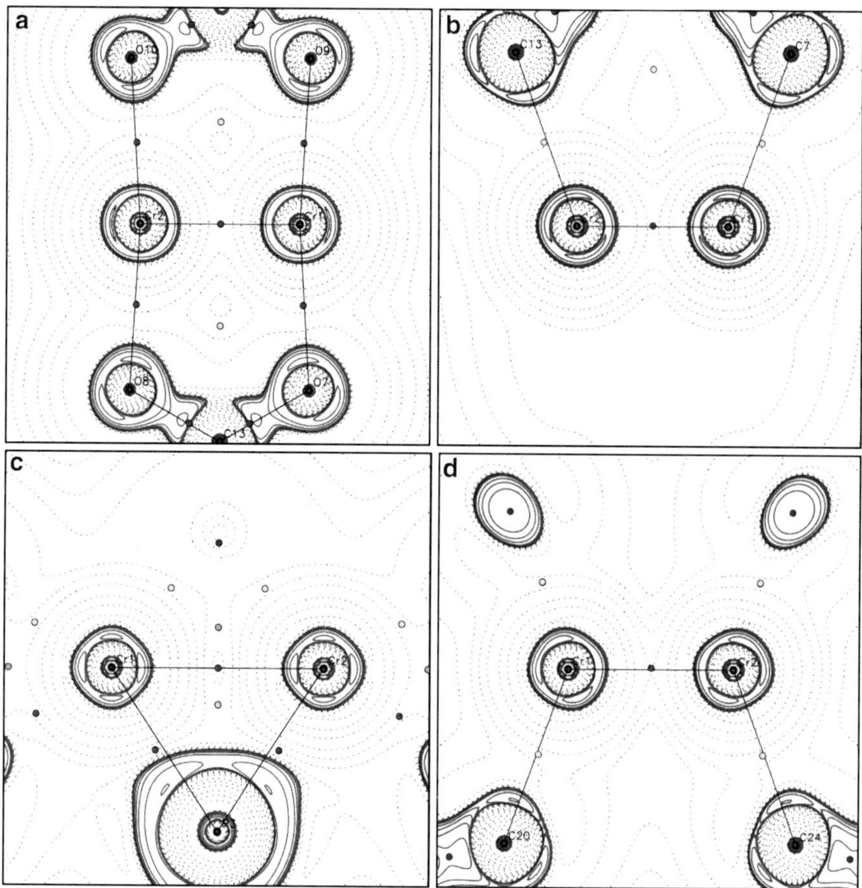

Fig. 9 Plots of the Laplacian $L \equiv -\nabla^2(\rho)$ in planes containing the M–M bond for complexes (**a**) $Cr_2(O_2CH)_4$, (**b**) $Cr_2(\eta^3\text{–}C_3H_5)_2(\mu\text{–}\eta^3\text{–}C_3H_5)_2$, (**c**) $[Cr_2Cp_2(\mu\text{–}\eta^5\text{–}P_5)]^-$ and (**d**) $Cr_2(\eta^4\text{–}C_8H_8)_2(\mu\text{–}\eta^8\text{–}C_8H_8)$. Positive contours as *solid lines* and negative contours as *broken lines*. Interatomic vectors are indicated, with critical points in $\rho(\mathbf{r})$ close to the plane

where the metal–ligand bonding is relatively ionic, there is only a marginal polarisation of the metal atoms, resulting in small, bond-opposed, charge concentrations. The situation is similar for $Cr_2(\eta^3–C_3H_5)_2(\mu–\eta^3–C_3H_5)_2$ (Fig. 9b), and here there are charge depletions along the Cr–Cr vector. In contrast, for $[Cr_2Cp_2(\mu–\eta^5–P_5)]^-$ (Fig. 9c), there are obvious charge concentrations lying along the Cr–Cr vector. The situation seen in Fig. 9b and also for $Cr_2(\eta^4–C_8H_8)_2(\mu–\eta^8–C_8H_8)$ (Fig. 9d), where charge depletions are seen along the M–M axis, was also previously observed for $Mn_2(CO)_{10}$ [65].

5 Conclusions

In this book chapter, we have discussed the connections between quantum mechanical observables (derived from electron density) and the metal–metal bond order (a concept that arises only from a molecular orbital solution to the Schrödinger equation). The data used for these discussions come partly from experimental electron density reconstruction using multipolar methods and partly from theoretical (DFT) wave function calculations. We can safely draw some conclusions, but some open problems remain and will probably continue to do so – meaning that no direct link is probably possible between the concept of bond order and actual observable properties.

We can certainly say that QTAIM studies outlined above suggest that the metal–metal bond in molecular compounds is a partial covalent bond (partial in the sense that not necessarily a whole pair of electrons are involved), somewhat similar to Pauling's original definition for the metallic bond [10]. Even in simple cases of formally single unsupported M–M bonds, some delocalisation involving the ancillary ligands is evident, making the distinction between "direct" and "through-bridge" bonding a very blurred one. M–M bonds present a much wider spectrum of properties than many other types of bond. More problematic is the analysis of multiple M–M bonds. In fact, evidence for multiple bonding interactions is not so easily extracted from the electron density. In some ways, this problem is not new. As is the case for the chemical bonds in organic molecules, it is not so easy to retrieve information on the bond order, unless resorting to empirical correlations. However, the strongly localised multiple bonds of C, O and N atoms typically give rise to large delocalisation indices, making clear the bond characteristics. Even for these atoms, as recently pointed out by Firme et al. [129], the relationship between formal bond order and the delocalisation indices is not transferable between atom types. The situation is much less straightforward with M–M bonds, because the inherently larger delocalisation of electrons involving ligand atoms makes $\delta(\Omega_M–\Omega_M)$ much smaller than the formal bond orders. Decomposition of the electron delocalisation indices into contributions from individual MOs could clarify the picture, but the purist would then argue that this is no longer an analysis of a pure quantum mechanical observable. As a side issue, the reader should be aware that $\delta(\Omega_M–\Omega_M)$ is most often computed at the DFT level, using the atomic overlap matrix decomposition, which actually gives a true observable only at the Hartree–Fock level.

Another way to visualise bonds and bond multiplicity comes from DAFH plots, which is, in a way, a visualisation of the contributions to the delocalisation indices. Domain natural orbitals, within a DAFH analysis, give an invariant representation that does not depend on the set of molecular orbitals adopted. This would give the possibility of ascertaining the presence of contributions coming from different types of electrons in the molecule. Nevertheless, a completely unambiguous indicator of bond order has not yet been proposed, or even an indicator that gives idea of a bond order larger than 1. Moreover, another problem in electron density analysis of M–M bonds is the difficult discrimination between covalent and non-covalent interactions. Although some obvious covalent M–M bonds have been inappropriately assigned as "closed-shell" interactions, we expect there to be problems when addressing truly non-covalent M–M bonds (as, for example, Ag···Ag or Au···Au interactions) using QTAIM methodology.

Despite our somewhat limiting conclusions, this chapter shows the kind of information that becomes available after accurate theoretical or experimental analysis of the electron density distribution. Possible progress in the field could come, for example, from more use of multiconfigurational wavefunctions, including relativistic corrections and spin–orbit coupling if necessary. Work mentioned above suggests this approach is already necessary for multiple M–M bonds [97, 98, 113, 116, 118, 122], and a recent report by Platts et al. on $Co_2(CO)_6(RC{\equiv}CR)$ complexes [128] implies this may also be true for single M–M bonded systems. Since the overwhelming number of theoretical QTAIM studies that have been currently undertaken on M–M bonded systems have relied on the (single configurational) DFT ansatz, one has to question whether some different conclusions could arise if more exact wavefunctions were used. Application of the interacting quantum atom approach or the electron number distribution function (both being developed by the group of Martín Pendás) could also improve the chemical interpretation of the (theoretical) electron density distribution.

From an experimental point of view, higher levels of accuracy in the regions of low electron density, which has produced much progress in the analysis of (even weaker) intermolecular interactions, should result in more confident interpretations also for M–M-bonded systems. Finally, we note that Matito and Solà [130] have recently reviewed the role of electron delocalisation in transition metal complexes, and have also discussed the use of the delocalisation index in understanding metal–metal bonding.

Acknowledgments PM thanks the Swiss National Science Foundation for Support (project 200021_125313).

References

1. Lewis GN (1916) J Am Chem Soc 38:762
2. Bader RFW (1990) Atoms in molecules: a quantum theory. Oxford University Press, Oxford
3. Bader RFW, Johnson S, Tang T-H, Popelier PLA (1996) J Phys Chem A 100:15398

4. Bader RFW, Stephens ME (1975) J Am Chem Soc 97:7391
5. Fradera X, Austen MA, Bader RFW (1999) J Phys Chem A 103:304
6. Cotton FA (1966) Q Rev (Chem Soc London) 20:389
7. Aromi G, Brechen EK (2006) Struct Bonding 122:1
8. Mingos DMP, Wales DJ (1990) Introduction to cluster chemistry. Prentice-Hall International, London
9. Mingos DMP, May AS (1990) In: Shriver DF, Kaesz HD, Adams RD (eds) The chemistry of metal cluster compounds. VCH, Weinheim, Germany, pp 11–119 (Chap. 2)
10. Pauling L (1948) The nature of the chemical bond. Cornell University Press, Ithaca
11. Anderson WP, Burdett JK, Czech PT (1994) J Am Chem Soc 116:8808
12. Allen LC, Capitani JF (1994) J Am Chem Soc 116:8810
13. Silvi B, Gatti C (2000) J Phys Chem A 104:947
14. Gatti C, Fantucci P, Pacchioni G (1987) Theor Chim Acta (Berlin) 72:433
15. Martin M, Rees B, Mitschler A (1982) Acta Cryst B38:6
16. Leung PC, Coppens P (1983) Acta Cryst B39:535
17. Mitschler A, Rees B, Lehmann MS (1978) J Am Chem Soc 100:3390
18. Wang Y, Coppens P (1976) Inorg Chem 15:1122
19. Clemente DA, Biagini MC, Rees B, Herrmann WA (1982) Inorg Chem 21:3741
20. Geibel W, Wilke G, Goddard R, Krüger C, Mynott R (1978) J Organomet Chem 160:139
21. Benard M, Coppens P, DeLucia ML, Stevens ED (1980) Inorg Chem 19:1924
22. Low AA, Kunze KL, MacDougall PJ, Hall MB (1991) Inorg Chem 30:1079
23. MacDougall PJ (1990) Ph. D. Thesis, McMaster University
24. Bader RFW (1998) J Phys Chem A 102:7314
25. Bader RFW (2009) J Phys Chem A 113:10391
26. Keith TA, Bader RFW, Aray Y (1996) Int J Quantum Chem 57:183
27. Farrugia LJ, Evans C, Lentz D, Roemer M (2009) J Am Chem Soc 131:1251
28. Pendás AM, Francisco E, Blanco MA, Gatti C (2007) Chem Eur J 13:9362
29. Jayatilaka D, Grimwood DJ (2001) Acta Cryst A57:76
30. Powell HM, Evans RVG (1939) J Chem Soc 286
31. Cotton FA, Troup JM (1974) J Chem Soc Dalton Trans 800
32. Greenwood NN, Earnshaw A (1984) Chemistry of the elements. Pergamon Press, Oxford, p 1282
33. Summerville RH, Hoffmann R (1979) J Am Chem Soc 79:1501
34. Heijser W, Baerends EJ, Ros P (1980) Faraday Symp 14:211
35. Bauschlicher CW (1986) J Chem Phys 84:872
36. Rosa A, Baerends E (1991) New J Chem 15:815
37. Bo C, Sarasa J-P, Poblet J-M (1993) J Phys Chem 97:6362
38. Reinhold J, Hunstock E, Mealli C (1994) New J Chem 18:465
39. Reinhold J, Kluge O, Mealli C (2007) Inorg Chem 46:7142
40. Ponec R, Cooper DL (2007) Faraday discuss 135:31
41. Ponec R, Lendvay G, Chaves J (2008) J Comput Chem 29:1387
42. Ponec R, Gatti C (2009) Inorg Chem 48:11024
43. Elschenbroich C (2006) Organometallics. Wiley-VCH, Weinheim, p 361
44. Macchi P, Donghi D, Sironi A (2005) J Am Chem Soc 127:16494
45. Overgaard J, Iversen BB, Palli SP, Timco GA, Gerbeleu NV, Larsen FK (2002) Chem Eur J 8:2775
46. Overgaard J, Larsen FK, Schiøtt B, Iversen BB (2003) J Am Chem Soc 125:11088
47. Poulsen RD, Bentien A, Graber T, Iversen BB (2004) Acta Cryst A60:382
48. Poulsen RD, Bentien A, Chevalier M, Iversen BB (2005) J Am Chem Soc 127:9156
49. Poulsen RD, Jørgensen MRV, Overgaard J, Larsen FK, Moergenroth W, Graber T, Chen Y-S, Iversen BB (2007) Chem Eur J 13:9775
50. Clausen HF, Overgaard J, Chen Y-S, Iversen BB (2008) J Am Chem Soc 130:7988
51. Overgaard J, Larsen FK, Timco GA, Iversen BB (2009) J.C.S. Dalton Trans 664

52. Pilet S, Souhassou M, Lecomte C, Rabu P, Drillon M, Massabrio C (2006) Phys Rev B73:115116
53. Pilet S, Souhassou M, Lecomte C (2004) Acta Cryst A60:455
54. Pilet S, Souhassou M, Mathoniére C, Lecomte C (2004) J Am Chem Soc 126:1219
55. Yasui M, Takayama R, Akiyama N, Hashizume D, Iwasaki F (2002) Mol Cryst Liq Cryst 376:519
56. Coppens P, Iversen BB, Larsen FK (2005) Coord Chem Rev 249:179
57. Farrugia LJ, Middlemiss DS, Sillanpää R, Seppälä P (2008) J Phys Chem A 112:9050
58. Jauch W, Reehuis M (2002) Phys Rev B 65:12511
59. Jauch W, Reehuis M (2003) Phys Rev B 67:184420
60. Jauch W, Reehuis M (2004) Phys Rev B 70:195121
61. Slater JC (1974) The self-consistent field for molecules and solids: quantum theory of molecules and solids, vol 4. McGraw-Hill, New York
62. MacDougall PJ, Hall MB (1990) Trans Am Cryst Assoc 26:105
63. Bianchi R, Gervasio G, Marabello D (1998) Chem Commun 1535
64. Bianchi R, Gervasio G, Marabello D (2000) Inorg Chem 39:2360
65. Farrugia LJ, Mallinson PR, Stewart B (2003) Acta Cryst B59:234
66. Brown DA, Chambers WJ, Fitzpatrick AJ, Rawlinson SRM (1971) J Chem Soc A 720
67. Rosa A, Ricciardi G, Baerendts EJ, Stufkens DJ (1995) Inorg Chem 34:3425
68. Gervasio G, Bianchi R, Marabello D (2004) Chem Phys Lett 387:481
69. Ponec R, Yuzhakov G, Sundberg MR (2005) J Comp Chem 26:447
70. Macchi P, Proserpio DM, Sironi A (1998) J Am Chem Soc 120:13429
71. Cremer D, Kraka E (1984) Croat Chem Acta 57:1259
72. Cremer D, Kraka E (1984) Angew Chem Int Ed 23:67
73. Novozhilova IV, Volkov AV, Coppens P (2003) J Am Chem Soc 125:1079
74. Kim CD, Pillet S, Wu G, Fullagar WK, Coppens P (2002) Acta Cryst A58:133–137
75. Macchi P, Sironi A (2003) Coord Chem Rev 238–239:383
76. Spackman MA, Maslen EN (1985) Acta Cryst A41:347
77. Berlin TZ (1951) Chem Phys 19:208
78. Macchi P, Garlaschelli L, Martinengo S, Sironi A (1999) J Am Chem Soc 121:10428
79. Macchi P, Garlaschelli L, Sironi A (2002) J Am Chem Soc 124:14173
80. Farrugia LJ, Evans C (2005) Comptes Rendus Chemie 8:1566
81. Flierler U, Burzler M, Leusser D, Henn J, Ott H, Braunschweig H, Stalke D (2008) Angew Chem Int Ed 47:4321
82. Götz K, Kaupp M, Braunschweig H, Stalke D (2009) Chem Eur J 15:623
83. Overgaard J, Clausen HF, Platts JA, Iversen BB (2008) J Am Chem Soc 130:3834
84. Overgaard J, Platts JA, Iversen BB (2009) Acta Cryst B65:715
85. Bianchi R, Gervasio G, Marabello D (2001) Acta Cryst B57:638
86. Bianchi R, Gervasio G, Marabello D (2001) Helv Chim Acta 84:722
87. Farrugia LJ (2005) Chem Phys Lett 414:122
88. Gervasio G, Bianchi R, Marabello D (2005) Chem Phys Lett 407:18
89. Macchi P, Sironi A (2001) XX European Crystallographic Meeting, Cracow, 26–31 August 2001
90. Van der Maelen JF, Gutiérrez-Puebla E, Monge A, García-Granda S, Resa I, Carmona E, Fernández-Diáz MT, McIntyre GJ, Pattison P, Weber H-P (2007) Acta Cryst B63:862
91. Overgaard J, Jones C, Stasch A, Iversen BB (2009) J Am Chem Soc 131:4208
92. Green SP, Jones C, Stasch A (2008) Angew Chem Int Ed 47:9079
93. Del Rio D, Galindo A, Resa I, Carmona E (2005) Angew Chem Int Ed 44:1244
94. Philpott MR, Kawazoe Y (2006) J Mol Struct THEOCHEM 773:43
95. Datta A (2008) J Chem Phys C 112:18727
96. Cotton FA, Murillo CA, Walton RA (eds) (2006) Multiple bonds between metal atoms. Springer, New York
97. Gagliardi L, Roos BO (2003) Inorg Chem 42:1599

98. Saito K, Nakao Y, Sat H, Sakaki S (2006) J Phys Chem A 110:9710
99. Krapp A, Lein M, Frenking G (2008) Theor Chem Acc 120:3131
100. Ponec R, Yuzhakov G (2007) Theor Chem Acc 118:791
101. Molina Molina J, Dobado JA, Heard GL, Bader RFW, Sundberg MR (2001) Theor Chem Acc 105:365
102. Ponec R, Yuzhakov G, Gironés X, Frenking G (2004) Organometallics 23:1790
103. Goh LY, Mak CW (1986) Chem Commun 1474
104. Adams RD, Collins DE, Cotton FA (1974) J Am Chem Soc 96:749
105. Nguyen T, Sutton AD, Brynda M, Fettinger JC, Long GL, Power PP (2005) Science 310:844
106. Kreisel KA, Yap GPA, Dmitrenko O, Landis CR, Theopold KH (2007) J Am Chem Soc 129:14162
107. Tsai Y-C, Hsu C-H, Yu J-SK, Lee G-H, Wang Y, Kuo T-S (2008) Angew Chem Int Ed 47:7250
108. Hsu C-H, Yu J-SK, Yen C-H, Lee G-H, Wang Y, Tsai Y-C (2008) Angew Chem Int Ed 47:9933
109. Noor A, Wagner FR, Kempe R (2008) Angew Chem Int Ed 47:7246
110. Tsai Y-C, Chen H-Z, Chang C-C, Yu J-SK, Lee G-H, Wang Y, Kuo T-S (2009) J Am Chem Soc 131:12534
111. Frenking G (2005) Science 310:796
112. Radius U, Breher F (2006) Angew Chem Int Ed 45:3006
113. Brynda M, Gagliardi L, Widmark P-O, Power PP, Roos BO (2006) Angew Chem Int Ed 45:3804
114. Merino G, Donald KJ, D'Acchioli JS, Hoffmann R (2007) J Am Chem Soc 129:15295
115. Roos BO, Borin AC, Gagliardi L (2007) Angew Chem Int Ed 46:1469
116. La Macchia G, Aquilante F, Veryazov V, Roos BO, Gagliardi L (2008) Inorg Chem 47:11455
117. DuPré DB (2009) J Phys Chem A 113:1559
118. Brynda M, Gagliardi L, Roos BO (2009) Chem Phys Lett 471:1
119. Wagner FW, Noor A, Kempe R (2009) Nat Chem 1:529
120. Ponec R, Feixas F (2009) J Phys Chem A 113:8394
121. Tsai Y-C, Chang C-C (2009) Chem Lett 38:1122
122. La Macchia G, Gagliardi L, Power PP, Brynda M (2008) J Am Chem Soc 130:5104
123. Wagner FR, Kohout M, Grin Y (2008) J Phys Chem A 112:9814
124. Gatti C, Lasi D (2007) Faraday Discuss 135:55
125. Bader RFW, Gatti C (1998) Chem Phys Lett 287:233
126. Gatti C, Cargnoni F, Bertini L (2003) J Comput Chem 24:422
127. Gatti C, Bertini L (2004) J Acta Cryst A60:438
128. Llusar R, Beltrán A, Andrés J, Fuster F, Silvi B (2001) J Phys Chem A 105:9460
129. Firme CL, Antunes OAC, Esteves PM (2009) Chem Phys Lett 468:129
130. Matito E, Solà M (2009) Coord Chem Rev 253:647
131. Platts JA, Evans GJS, Coogan MP, Overgaard J (2007) Inorg Chem 46:6291
132. Ortin Y, Lugan N, Pillet S, Souhassou M, Lecomte C, Costaus K, Saillard J-Y (2005) Inorg Chem 44:9607
133. Poulsen RD, Overgaard J, Schulman A, Østergaard C, Murillo CA, Spackman MA, Iversen BB (2009) J Am Chem Soc 131:7580
134. Bo C, Costas M, Poblet JM, Rohmer M-M, Benard M (1996) Inorg Chem 35:3298

Struct Bond (2012) 146: 159–208
DOI: 10.1007/430_2012_77
© Springer-Verlag Berlin Heidelberg 2012
Published online: 27 March 2012

On the Nature of β-Agostic Interactions: A Comparison Between the Molecular Orbital and Charge Density Picture

Wolfgang Scherer, Verena Herz, and Christoph Hauf

Abstract The phenomenon and nature of agostic interactions are reviewed in light of combined molecular orbital and charge density studies. As an introduction a historical perspective is given, illustrating the successes and short falls of the various bonding concepts developed during the past 45 years since the discovery of the phenomenon in transition metal complexes. The finding that β-agostic species might represent stable intermediates along the β-elimination reaction co-ordinate classifies them as suitable benchmark systems to study the microscopic origin of C–H bond activation processes. We outline the salient electronic parameters that control and quantify the extent of agostic interactions on the basis of physically observable charge density properties. Despite the focus on charge density studies, we also complement these studies with arguments based on molecular orbital theory and an irrefutable body of crystallographic, kinetic, and spectroscopic evidence.

Keywords Agostic interaction · Bond activation · Charge density studies · Computational studies · DFT · Electron delocalization · Molecular orbitals · Multipolar refinements · Negative hyperconjugation · Topological analysis · X-ray analysis

Contents

W. Scherer (✉), V. Herz, and C. Hauf
Institute of Physics, University of Augsburg, Universitätsstr. 2, 86159 Augsburg, Germany
e-mail: Wolfgang.scherer@physik.uni-augsburg.de

1 Introduction

In general, C–H bonds can be considered chemically inert as a result of their strength, nonpolar nature, and low polarizability. However, over half a century ago, Burawoy [1] along with Pitzer and Gutowsky [2] proposed the dimeric structure of $[Me_3Al]_2$ to result from Al···H–C bridging interactions. Throughout the 1960s and 1970s, a growing number of crystallographic and spectroscopic studies suggested that transition metals were capable of forming significant interactions with the C–H bonds of their appended ligands. The first such report came in 1965 from La Placa and Ibers, who provided evidence of the close approach of an *ortho*-C–H bond of a triphenylphosphine ligand to the Ru(d^6) center in $[RuCl_2(PPh_3)_3]$ [3].[1] However, the relevance of this new type of interaction was only realized toward the end of the decade by Trofimenko. In his pioneering studies, he reported on a "hydridic" character in the NMR properties of the methylene groups in a series of transition metal–pyrazolylborato complexes [5–7], concluding *hydrogens are intruding into empty metal orbitals* – a concept later developed and refined by Brookhart and Green (BG) by coining the expression *agostic* interactions in 1983 [4, 8].[2]

1.1 Phenomenological Description of Agostic Interactions

According to the definition of BG, the term *agostic* is used to discuss the various manifestations of covalent interactions between C–H groups and transition metal

[1] It is interesting to note that the agostic interaction in this agostic benchmark complex represents a rare example of a so-called γ-agostic [4] interaction. In agostic alkyl complexes of early transition d^0 metal complexes, β-agostic interactions are generally stronger than their α- or γ-counterparts. Hence, the first literature example of an agostic transition metal complex already suggests that the nature and strength of an agostic interaction might depend on the electronic situation at the metal center (d-electron count) and that of the ligand (presence of hetero atoms).

[2] The term *agostic* has been introduced by MLH Green and is derived from the Greek word ἀγοστόζ, which might be translated as to clasp, to draw towards, to hold to oneself; see p. 3. of [8].

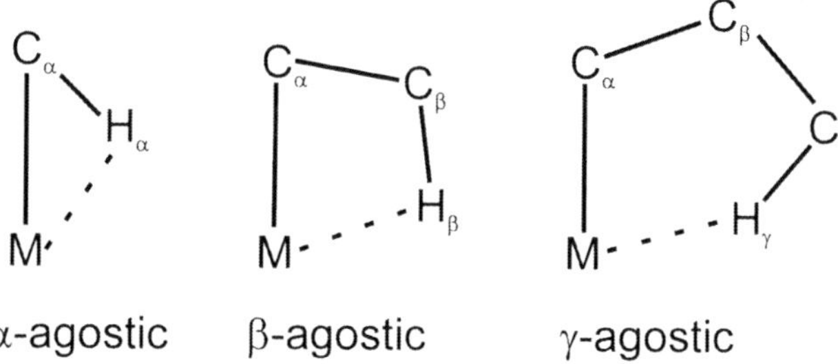

Scheme 1 Nomenclature of agostic interactions

centers in organometallic compounds. In these instances, the *hydrogen atom is covalently bonded simultaneously to both a carbon and to a transition metal atom* [8] (see Scheme 1 for the notation of agostic interactions introduced by BG).

Nowadays, this rather stringent original definition of agostic interactions has been extended to include non-covalent $M\cdots H_\beta{-}C_\beta$ interactions involving main group elements, such as Li, and polar H–X bonds (e.g., X = Si, Ge,...) or even C–C single bonds [9]. Hence, the meaning of the concept – a hitherto rare covalent interaction between a "chemically inert" C–H bond and transition metal centers – has been somewhat lost through its current usage. Scherer and McGrady (SMG) have therefore recently proposed a general phenomenological definition: *agostic interactions are characterized by the distortion of an organometallic moiety which brings an appended C–H bond into close proximity with the metal centre* [10]. Such a definition accommodates most of the examples reported in the literature but separates the nature of the phenomenon and the driving force behind it from its observable chemical consequence. Therefore, this phenomenological definition still needs to be complemented by a unifying bonding concept which promises systematic control of the driving forces of C–H activation in agostic complexes of early and late transition metal complexes.

Such a concept was introduced in 2003 on the basis of theoretical and experimental charge density analyses [11]. In this study, SMG showed that the interplay between *locally* induced sites of increased Lewis acidity and an alkyl ligand is crucial to the development of a β-agostic interaction in d^0 metal alkyls, which is driven by hyperconjugative delocalization of the σ(M–C) bonding electrons over the alkyl backbone. SMG also showed that C–H bond activation due to a covalent M···H–C interaction typically plays only a secondary role in case of d^0 complexes. Accordingly, the bonding between the metal atom and the agostic ethyl or amido groups is effectively established by *one* electron pair/*one* molecular orbital, in agreement with earlier findings [11–16]. This bonding scenario conforms with the puzzling observations based on neutron diffraction data in which agostic C–H bonds may not be elongated even by coordination to Lewis-acidic lanthanide metal centers, such as $[Cp^*Y(OC_6H_3{}^tBu_2)CH(SiMe_3)_2]$ and $[Cp^*La\{CH(SiMe_3)_2\}_2]$ [17] or in so-called poly-agostic $[Nd(AlMe_4)_3]$ [18]. Therefore, it is useful to distinguish these M···H–C interactions from those displayed by late transition metals where the possibility of significant back donation from the electron-rich transition metal center to the ligand leads to pronounced C–H bond elongation >0.1 Å and the establishment of M···H bond paths (BP) in the corresponding electron density maps [10].

1.2 Early Bonding Concepts and Their Pitfalls

BG initially proposed in their comprehensive review on agostic interactions in 1983 [4] and 1988 [8] that the agostic bonding should be considered as three-center, two-electron (3c2e) covalent bond, with *donation of C–H bonding electrons into a vacant atomic orbital on the transition metal atom* (Scheme 2, left). This statement conforms with the earlier suggestion from Trofimenko [5] in 1967 stating that the agostic *hydrogens are intruding into empty metal orbitals.* Also Cotton classified already in 1974 the agostic interaction in $[Mo\{Et_2B(pz)_2\}(\eta^3\text{-}C_3H_4Ph)(CO)_2]$ as *a three-center, two-electron bond encompassing the C···H···Mo atoms* [19] in analogy with bonding concepts in borane chemistry. He further drew comparisons with the concept in organolithium chemistry accounting for *three center bonds of the form C···H···Li* [20]. Kaufmann et al. finally extended this concept to enclose also main group alkyls with the suggestion that the *σCH–Li interaction is the organolithium form of the agostic interaction* [21].

In 1982, Green et al. reported the d^0 titanium alkyl complexes $[RTiCl_3(dmpe)]$ (dmpe = $Me_2PCH_2CH_2PMe_2$; R = Me **1** or Et **2**, see Fig. 1) [22, 23] representing textbook examples of M···H–C α and β agostic interactions, respectively. Green et al. concluded that in **2** *the ethyl group models a stage about half-way along the reaction coordinate for a β-elimination reaction* (Scheme 3). However, the elimination product would be unstable "since the d^0 titanium center cannot formally back donate electrons to the ethylene ligand," consistent with the general absence of β-elimination in the chemistry of d^0 transition metal alkyls [23]. The finding that agostic species might represent stable intermediates along the β-elimination reaction coordinate and can be the *ground state* under appropriate conditions prompted a major shift in paradigm in organometallic chemistry and catalysis [22, 23]. Furthermore, at this early stage it was already clear that agostic d^0 complexes should be distinguished from their later d^n ($n \geq 2$) counterparts, which are able to

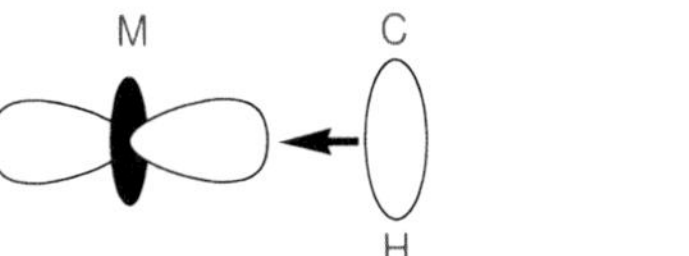

Scheme 2

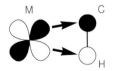

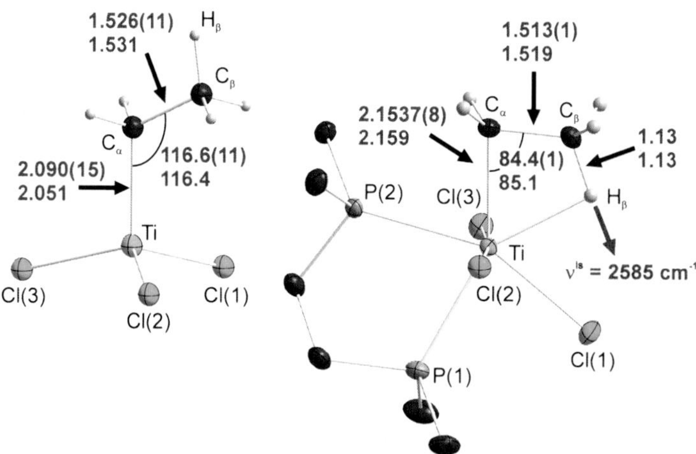

Fig. 1 Molecular models of the d^0 titanium alkyl complex $EtTiCl_3$ **3** (gas electron diffraction model [12]) and its corresponding phosphine adduct $EtTiCl_3(dmpe)$ **2** (high-resolution X-ray model at 100 K; 50% probability level [11]). Salient bond distances (Å) and angles (°); theoretical values (*blue color*) are given below the experimental ones (*red color*)

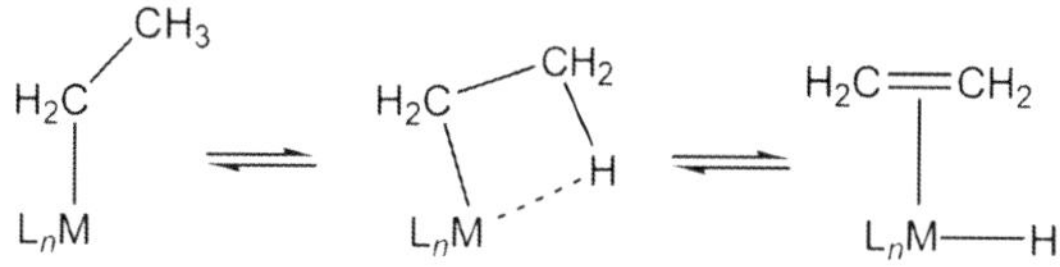

$H_2C(CH_3)$–L_nM ⇌ H_2C–CH_2···H–L_nM ⇌ $H_2C{=}CH_2$ / L_nM–H

Scheme 3

back donate electrons (Scheme 2) and support a nascent olefin ligand. Crabtree and Hamilton therefore compared agostic M···H–C interactions with the bonding situation in η^2-H_2 metal complexes and concluded in 1988 *instead of being strictly side-on as in the case of* η^2-H_2 *complexes, the C–H bond is usually skewed in such a way that the H atom is closer to the metal* [24]. They also tried to explain the relatively small lengthening of the agostic C–H bonds (typically smaller than 0.1 Å) relative to metal-activated η^2-H_2 bonds (ca. 0.2 Å): "Since we expect $M(d_\pi)$ to X–H(σ^*) backbonding" (X = C or H) "to be the chief factor affecting the X–H bond lengthening, the explanation may be that the C–H σ^* is less accessible than the corresponding H–H σ^*."

On the basis of the BG bonding concept which relies on the presence of dominant M ← H–C donation of electron density, the following control parameters for agostic interactions were elaborated: (1) the valence electron (VE) count displayed by the transition metal M should be equal or smaller than 16 VE; (2) a high Lewis acidity/positive charge at the metal atom is needed; (3) the degree of steric congestion at M measured primarily by its coordination number (CN) should be low; and (4) the presence of an available acceptor orbital of suitable symmetry at

M is essential [4, 8]. While these factors may be desirable prerequisites, they are in no way sufficient as a reliable basis for prediction of the phenomenon. In fact, the majority of organometallic complexes which formally satisfy conditions (1)–(4) show little or no evidence of agostic behavior. This maybe best illustrated in case of the four-coordinate (8 VE) species, $EtTiCl_3$ **3** and its (12 VE) dmpe adduct **2** (Fig. 1). Apparently, only the donor-stabilized complex **2** is characterized by pronounced agostic interactions, while **3** displays a rather undistorted ethyl group. This result is thus in clear conflict with the predictions by the BG model since **3** lacks any significant Ti···H–C interactions despite the presence of suitable d-acceptor orbitals at the metal and its lower VE count, higher Lewis acidity, and lower steric congestion in comparison with **2**. We will therefore outline at a later stage (Sects. 4 and 5) that in contrast to the BG model, (1) the presence and extent of *local Lewis-acidic sites* in the valence shell of the metal atom and (2) the *negative hyperconjugative delocalization of the M–C bonding electron pair* are the chief control parameters of β-agostic interactions in d^0 configurated transition metal alkyls.

It is interesting to note that the textbook example for a β-agostic interaction $EtTiCl_3$(dmpe) only displays a subtle *C–H bond elongation* of approximately 0.03 Å relative to a standard C(sp^3)–H bond in accordance with an *isolated stretching frequency* of 2,585 cm^{-1} [16]. Besides these fundamental criteria, the β-agostic interaction is clearly indicated by an acute ∠TiCC angle of 84.4(1)° and a short C–C bond length of 1.513(1) Å (reduced by ca. 0.03 Å relative to a standard C–C single bond length) revealing the partial olefinic character of **2**. However, in contrast to the BG bonding concept, the agostic proton does not reveal any "hydridic" character and displays a downfield chemical shift in the proton NMR (see Sect. 5.3 for a detailed discussion). This scenario changes dramatically, in case of the Spencer-type complex [25] $[(DCpH)Ni(d^tbpe)]^+[BF_4]^-$ ([DCpH = dicyclopentenyl) **5d** which can be obtained from the corresponding nickel olefin species **4d** upon protonation with HBF_4 [26] (Fig. 2). In this d^8 agostic benchmark system, we notice that all characteristic criteria suggest the presence of a stronger agostic interaction in $[(DCpH)Ni(d^tbpe)]^+[BF_4]^-$ **5d** relative to its d^0 congener **2**. Accordingly, we observe the expected upfield shift of the agostic proton ($\delta(^1H)$ = −5.37 ppm). We note, however, that both agostic protons in **2** and **5d** are not carrying any significant negative "hydridic" charge as imposed by the BG model (Tables 1–3).

It is therefore obvious that a concept is needed which goes beyond the BG model to understand the true nature of this unique interaction in organometallic chemistry. It will therefore be the goal of this contribution to analyze the nature of agostic interactions by experimental and theoretical charge density methods to elaborate reliable criteria for the prediction and control of β-agostic interactions in early and late transition metal complexes. If not specified otherwise, all theoretical values were obtained by DFT calculations [31, 32] using the BP86 functional [33, 34] in combination with the basis set of triple-ζ quality (TZ2P) and the spin-orbit ZORA-Hamiltonian [35] in the following.

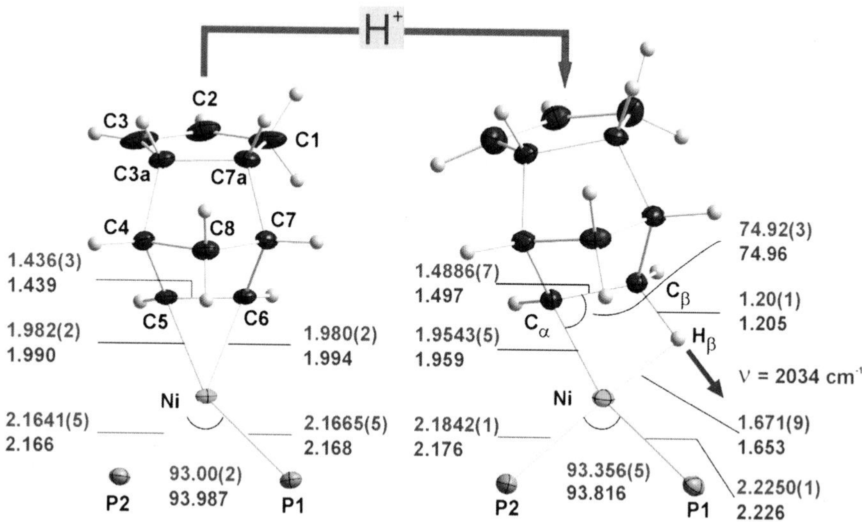

Fig. 2 ORTEP representations (50% probability level) of the d^{10} nickel olefin complex (DCp)Ni (d'bpe) **4d** and its protonated agostic d^8 cation $[(DCpH)Ni(d'bpe)]^+[BF_4]^-$ **5d** at 100 K. Salient bond distances (Å) and angles (°); theoretical values (*blue color*) are given below the experimental ones (*red color*)

Table 1 Comparison of the experimental and calculated ^1H NMR chemical shifts (δ) (ppm), shielding contributions σ, and atomic charges Q^{H}_{AIM} of the agostic protons in **2**, **6**, and **5d**, respectively

	$\angle MC_\alpha C_\beta$ (°)	d M–H_β (Å)	d C_β–H_β (Å)	δ (ppm)	σ^d (ppm)	σ^p (ppm)	σ^{so} (ppm)	Q^H_{AIM} (e)
5d$_{exp}$	74.92(3)	1.671(9)	1.20(1)	−5.27/−5.46[a]	–	–	–	−0.01
5d$_{calc}$	74.93	1.653	1.205 ↑	−5.62 ↑	28.43	6.09	2.69	−0.045
6$_{calc}$	84.4	2.029	1.156	−1.30	29.25	4.30	−0.66	−0.103
2$_{exp}$	84.4(1)	2.096	1.13	2.7[b]	–	–	–	0.13
2$_{calc}$	85.1	2.110	1.131	5.10	27.64	−1.08	−0.08	−0.031

[a]Two diastereomers in solution

[b]Averaged signal due to methyl group rotation; the isotopic perturbation of resonance (IPR) method of Shapley et al. [27] yields a negative IPR value in agreement with the deshielding of the agostic proton [15]

2 Characterization of Agostic Interactions by Charge Density Analyses

2.1 *Bond Path Analysis of Agostic Moieties*

2.1.1 Are Agostic Interactions Reflected by a M···H Bond Path?

In the framework of the theory of atoms in molecules, Bader concluded that *atoms linked by a line along which the density is a maximum with respect to any neighbouring line, a bond path, are bonded to one another and share a common*

Table 2 Comparison of experimental and calculated ^{1}H NMR chemical shifts (δ) (ppm), shielding contributions σ, and atomic charges Q^H_{AIM} of the agostic protons in Ti alkyl (X_α=C) and amido (X_α=N) benchmark complexes

	$\angle TiX_\alpha C_\beta$ (°)	d Ti–H (Å)	δ (ppm)	σ^d (ppm)	σ^p (ppm)	σ^{so} (ppm)	Q^H_{AIM} (e)
2	84.4(1)[a]	2.096[a]	2.7[b,c]	–	–	–	0.13
BP86	85.1	2.110	4.92	27.89	−1.50	0.04	−0.031
PBE0			5.10	27.64	−1.08	−0.08	
22	–	–	–	–	–	–	–
BP86	84.4	2.073	−0.33	28.41	3.33	−0.06	−0.071
PBE0			−0.02	29.28	3.52	−0.20	
6	–	–	–	–	–	–	–
BP86	84.4	2.029	−1.77	28.32	5.20	−0.40	−0.103
PBE0			−1.30	29.25	4.30	−0.66	
17	100.4(1)	2.25(3)	6.77[d]	–	–	–	–
BP86	101.0	2.323	6.46	29.25	−4.43	0.06	−0.003
PBE0			6.60	30.31	−5.38	0.05	
23a[e]	117.4(2)	2.80(2)	5.69	–	–	–	–
PBE0			5.59	29.08	−3.10	0.02	
23b[f]	114.2(2)	2.67(3)	5.92	–	–	–	–
PBE0			5.54	28.99	−3.09	0.15	
23c[f]	104.5(1)	2.38(2)	6.57	–	–	–	–
PBE0			6.55	29.79	−4.75	0.00	

[a]Reference [11]
[b]At 173 K [15]
[c]Averaged signal due to methyl group rotation
[d]At 178 K in toluene-d_8
[e]Reference [28]
[f]Reference [29]

interatomic surface (IAS) [36, 37]. Accordingly, AIM theory should provide a clear criterion for the presence or absence of an agostic interaction involving a metal and a hydrogen atom of an appended ligand. Hence in a pioneering study, Popelier and Logothetis analyzed the topology of theoretical electron density distributions of the agostic model complexes $CH_3TiCl_2^+$, $C_2H_5TiCl_2^+$, and $C_3H_7TiCl_2^+$ and suggested: *The existence of an agostic bond is clearly proven by the following triplet of concomitant topological objects: a bond critical point (BCP), a BP and an IAS* [38]. However, they also noticed a close proximity of the ring critical point (RCP) in the β- and γ-agostic moieties to the Ti···H BCP which *indicates structural instability* and related *the ease with which the agostic bond can be ruptured* to the *conjuncture that this is a weaker bond*. Indeed, the first experimental charge density study of an agostic compound showed that even the textbook example of a β-agostic interaction **2** was characterized by a scenario in which the Ti···H BCP and the RCP inside the $MC_\alpha C_\beta H_\beta$ moiety appear to coalesce and annihilate each other (Fig. 3) causing a disrupture of the Ti···H BP [13].

Table 3 Experimental and calculated ^{1}H NMR chemical shifts (δ) (ppm) together with the diamagnetic (σ^d), paramagnetic (σ^p), and spin–orbit (σ^{so}) shielding contributions of the agostic protons in the cations **5a–5d**

	$\angle NiC_\alpha C_\beta$ (°)	d Ni–H (Å)	δ (ppm)	σ^d (ppm)	σ^p (ppm)	σ^{so} (ppm)	Q^H_{AIM} (e)
5a[a]	74.5(3)	1.64(2)	−5.75[b]	–	–	–	–
BP86	75.0	1.634	−3.04	26.49	7.06	0.839	−0.052
PBE0			−6.06	28.21	7.28	2.16	
5b	74.2(1)	1.64(4)	−5.05	–	–	–	–
BP86	74.9	1.653	−1.75	27.06	5.22	0.820	−0.045
PBE0			−4.86	28.74	5.40	2.31	
5c	74.60(12)	1.72(3)	−5.38	–	–	–	–
BP86	74.9	1.653	−2.17	27.02	5.65	0.84	−0.057
PBE0			−5.30	28.40	6.11	2.39	
5d	74.92(3)	1.671(9)	−5.27 −5.46[c]	–	–	–	0.001
BP86	75.0	1.653	−2.67	27.12	5.51	1.394	−0.045
PBE0			−5.62	28.43	6.09	2.69	

[a]Reference [30]
[b]At 173 K
[c]Two diastereomers in solution

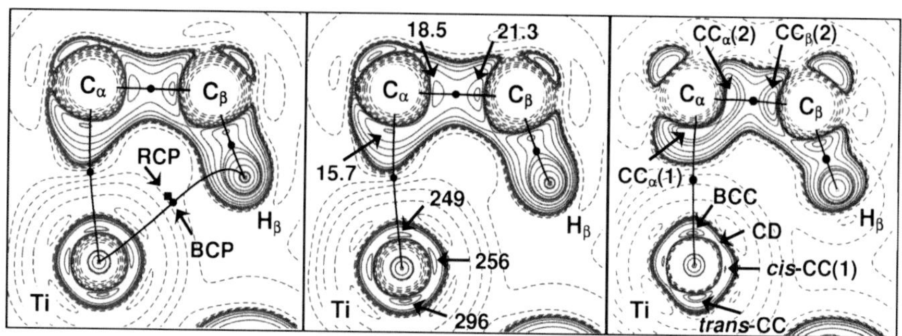

Fig. 3 Theoretical (*left*, *middle*) and experimental (*right*) $L(\mathbf{r}) = -\nabla^2\rho(\mathbf{r})$ contour maps and bond paths (*black solid line*) in the agostic $TiC_\alpha C_\beta H_\beta$ moiety of $EtTiCl_3$(dmpe) **2**. Positive (*solid*) and negative (*dashed*) contour lines are drawn at 0, $\pm2.0 \times 10^n$, $\pm4.0 \times 10^n$, $\pm8.0 \times 10^n$ eÅ^{-5} with $n = \pm3, \pm2, \pm1, 0$; one contour level deleted (200 eÅ^{-5}); extra level at 15, 25, 220, and 280 eÅ^{-5}. Salient values of the CPs in the $L(\mathbf{r})$ maps are specified in (eÅ^{-5}) (*middle*) and labeled (*right*) according to the notation given in Sect. 4.2. BCPs and RCPs are denoted by *filled circles* and *squares*, respectively. Note that the $Ti\cdots H_\beta$ bond path is not present in the optimized theoretical (*middle*) and experimental (*right*) model. The presence of a $Ti\cdots H_\beta$ interaction line was, however, enforced by a minute shortening of the Ti···H distance by 0.014 Å (*left*) to illustrate that **2** represents a system at the borderline of a so-called bond catastrophe [39, 40]. See Table 4 for a comparison of salient topological parameters of **2** (theory vs. experiment)

However, we will outline below that the bond catastrophe scenario and the missing Ti···$H_β$ BP in **2** is *not necessarily* a signature of the weakness of an agostic interaction but rather an indicator of its *delocalized* nature [10, 11]. Indeed, the ethyl group in **2** shows a significant structural distortion in comparison with the non-agostic reference system $EtTiCl_3$ (Fig. 1) as demonstrated by quite different ∠TiCC angles of 84.4(1) and 116.6(11), respectively. Also the pronounced C–H bond activation in **2** which is reflected by a remarkably low isolated stretching frequency $\nu(C_β–H_β)$ of 2,585 cm^{-1} highlights the chemical relevance of the β-agostic interaction in **2**. Accordingly, **2** might be classified as a model system at the early stage of the β-hydrogen elimination process (Scheme 3) – an important reaction channel in many organometallic transformations. A highly related situation was observed in case of the d^2 niobium complex $Tp^{Me2}NbCl(MeCCMe)(^iPr)$ (Tp^{Me2} = hydrido(3,5-dimethylpyrazolyl)borate) which represents one of the few examples where both α- and β-agostic conformations have been characterized in the same complex [41, 42]. In case of the α-agostic structure, no Nb···$H_α$ BCP could be derived irrespective of the computation level used. The general lack of M···$H_α$ BCPs is consistent with other theoretical charge density studies of α-agostic species [38, 43, 44]. The Nb···$H_β$ BCP was, however, found to be – like in **2** – close to the RCP of the β-agostic moiety, and its presence also depends on *very small changes in geometry or indeed methodology* which *can cause them to merge into a singularity* [41]. As an alternative analysis of the delocalization indices [45, 46][3] of the Ti···$H_β$, δ (Tables 4 and 5) might provide a more robust charge density criterion to classify (agostic) M···H bonds [48, 49] since it does not depend on the presence of a bond path [50].[4] Indeed, when agostic interactions are compared in late and early transition metal complexes, they nicely reflect the trend to a more pronounced covalent M···H bonding in the latter case; e.g., $\delta(Ti\cdots H_β)$ = 0.084 (in **2**), 0.158 (in $EtTiCl_2^+$ **6**), and 0.285 (in **5d**; Tables 4 and 5). However, Ti···H delocalization indices δ are only accessible by theoretical charge density analyses and do not provide a microscopic insight in the true nature and control parameters of the agostic interactions.

Indeed, in Sect. 4 we will outline that *negative hyperconjugative electron delocalization* involving the whole agostic alkyl backbone is the dominant driving force of agostic interactions in such early transition metal alkyls. Hence, in contrast to 3c4e hydrogen bonding, the phenomenon of agostic interactions is more complex and cannot be treated by analyzing the local electronic situation of the M···$H_βC_β$

---

[3] According to Poater et al., the delocalization indices $\delta(\Omega, \Omega')$ were calculated from the DFT wavefunctions using an approximate formula that makes use of an HF-like second order exchange density matrix. According to a recent study by Gatti et al. [46], this approximation affords $\delta(\Omega, \Omega')$ values which are very close to the HF ones if the HF and DFT optimized geometries are similar, although it erroneously implies that the electron pair density matrix can be constructed, within DFT, using the same simple formalism valid for the HF method.

[4] Delocalization indices were computed using GTO-type bases of triple-zeta quality and the B3LYP hybrid functional as implemented in Gaussian03 [50].

Table 4 Selected topological parameters $\rho(\mathbf{r})$ ($e\text{Å}^{-3}$), $\nabla^2\rho(\mathbf{r})$ ($e\text{Å}^{-5}$), ε, $H(\mathbf{r})$ (hartree Å^{-3}) and $G(\mathbf{r})/\rho(\mathbf{r})$ (hartree e^{-1}) and delocalization indices, δ, at the bond and ring critical points in the $[MC_\alpha C_\beta(H_\beta)]$ moieties of $EtTiCl_3(dmpe)$ **2** and $[EtTiCl_2]^+$ **6**, $[(\eta^2\text{-}C_2H_4)TiCl_2]$ **7**, $CpTiN(CHMe_2)_2Cl_2$ **17**, and $[EtTiCl_2{\cdot}PMe_3]^+$ **22**

		r	$\rho(\mathbf{r}_c)$	$\nabla^2\rho(\mathbf{r}_c)$	ε	$H(\mathbf{r}_c)$	$G(\mathbf{r}_c)/\rho(\mathbf{r}_c)$	δ
2								
$Ti\text{–}C_\alpha$	DFT	2.159	0.624	1.3	0.06	−0.213	0.490	0.705
	Exp.	2.1537(8)	0.500(9)	6.1(1)	0.22	−0.111	1.076	–
$C_\alpha\text{–}C_\beta$	DFT	1.519	1.593	−11.5	0.1	−1.257	0.283	1.034
	Exp.	1.513(1)	1.77(3)	−12.1(1)	0.10	−2.365	0.857	–
$C_\beta\text{–}H_\beta$	DFT	1.131	1.684	−18.7	0.02	−1.660	0.207	0.862
	Exp.	1.13	1.54(5)	−10.2(1)	0.14	−1.889	0.763	–
$H_\beta\text{–}Ti$[a]	DFT	2.114	–	–	–	–	–	0.084
	Exp.	2.096	–	–	–	–	–	–
6 eclipsed[b]								
$Ti\text{–}C_\alpha$	DFT	2.010	0.857	0.7	0.01	−0.385	0.507	0.894
$C_\alpha\text{–}C_\beta$	DFT	1.519	1.582	−11.2	0.11	−1.250	0.295	1.050
$C_\beta\text{–}H_\beta$	DFT	1.156	1.524	−14.6	0.03	−1.409	0.252	0.862
$H_\beta\text{–}Ti$	DFT	2.029	0.312	3.3	1.03	−0.031	0.833	0.158
RCP	DFT	–	0.304	4.4	–	−0.013	1.058	–
6 staggered[b,c]								
$Ti\text{–}C_\alpha$	DFT	1.996	0.877	0.8	0.05	−0.400	0.523	0.921
$C_\alpha\text{–}C_\beta$	DFT	1.546	1.495	−9.7	0.14	−1.120	0.296	1.009
$C_\beta\text{–}H_\beta'$	DFT	1.101	1.821	−23.9	0	−1.949	0.150	0.881
7								
Ti–C1	DFT	2.043	0.754	2.1	0.22	−0.300	0.595	0.769
C1–C2	DFT	1.486	1.673	−11.8	0.25	−1.399	0.341	1.205
C1–H2	DFT	1.095	1.835	−23.2	0.04	−1.957	0.179	0.962
RCP	DFT	–	0.599	8.2	–	−0.133	1.185	–
17								
$Ti\text{–}N_\alpha$	DFT	1.903	0.892	9.5	0.41	−0.302	1.086	0.992
$N_\alpha\text{–}C_\beta$	DFT	1.497	1.638	−11.1	0.02	−1.451	0.402	0.930
$C_\beta\text{–}H_\beta$	DFT	1.106	1.869	−23.8	0.00	−1.977	0.165	0.860
$H_\beta\text{–}Ti$	DFT	2.323	–	–	–	–	–	0.032
22								
$Ti\text{–}C_\alpha$	DFT	2.045	0.788	1.2	0.01	−0.327	0.526	0.816
$C_\alpha\text{–}C_\beta$	DFT	1.529	1.547	−10.7	0.11	−1.194	0.290	1.034
$C_\beta\text{–}H_\beta$	DFT	1.139	1.620	−17.0	0.03	−1.561	0.229	0.872
$H_\beta\text{–}Ti$	DFT	2.073	0.277	3.2	1.85	−0.018	0.874	0.120
RCP	DFT	–	0.274	3.9	–	−0.005	0.985	–

[a]No BCP point could be identified in the fully optimized geometries of **2** (Fig. 3)
[b]The labels "eclipsed" and "staggered" specify the ethyl group conformation
[c]H_β located in the symmetry plane is denoted by (′)

moieties alone. Furthermore, we will demonstrate in Sect. 4.3.2 that in extreme cases $M\cdots H_\beta C_\beta$ interaction might just play the role of secondary, closed shell interactions which solely assist the agostic interaction. As a consequence, the geometry and charge density features of agostic compounds usually do not change significantly during methyl group rotation as exemplified by the model system $EtTiCl_2^+$ **6**

Table 5 Selected topological parameters $\rho(\mathbf{r})$ (eÅ^{-3}), $\nabla^2\rho(\mathbf{r})$ (eÅ^{-5}), ε, $H(\mathbf{r})$ (hartree Å^{-3}) and $G(\mathbf{r})/\rho(\mathbf{r})$ (hartree e^{-1}) and delocalization indices, δ, at the bond and ring critical points in the [NiC_2] fragment of the olefin complexes (η^2-C_2H_4)Ni(d′bpe) **4a** and (DCp)Ni(d′bpe) **4d**. Corresponding experimental and theoretical values in the agostic [$NiC_\alpha C_\beta H_\beta$] moieties of the protonated nickel olefin compounds [(η^2-C_2H_5) Ni(d′bpe)]$^+$ **5a** and [(DCpH)Ni(d′bpe)]$^+$[BF_4]$^-$ **5d**

		r	$\rho(\mathbf{r}_c)$	$\nabla^2\rho(\mathbf{r}_c)$	ε	$H(\mathbf{r}_c)$	$G(\mathbf{r}_c)/\rho(\mathbf{r}_c)$[a]	δ
4a								
Ni–C_α	DFT	1.962	0.705	5.6	0.90	−0.251	0.910	0.692
	Exp.	1.9708(4)	0.671(6)	8.4(1)	1.03	−0.217	1.201	–
C_α–C_β	DFT	1.422	1.922	−16.3	0.26	−1.827	0.356	1.275
	Exp.	1.4189(6)	2.107(9)	−20.9(1)	0.25	−3.272	0.859	–
C_β–N_i	DFT	1.962	0.704	5.6	0.91	−0.251	0.910	0.692
	Exp.	1.9715(4)	0.670(2)	8.4(1)	1.05	−0.216	1.201	–
RCP	DFT	–	0.668	8.2	–	−0.193	1.145	–
	Exp.	–	0.648	8.9	–	−0.82	1.243	–
5a								
Ni–C_α	DFT	1.934	0.771	4.3	0.21	−0.314	0.795	0.808
C_α–C_β	DFT	1.478	1.743	−13.7	0.14	−1.500	0.309	1.083
C_β–H_β	DFT	1.215	1.351	−10.7	0.07	−1.099	0.257	0.734
H_β–Ni	DFT	1.634	0.592	6.6	1.02	−0.189	1.075	0.297
RCP	DFT		0.532	6.8	–	−0.117	1.109	–
4d								
Ni–C1	DFT	1.994	0.683	5.3	0.83	−0.238	0.896	0.666
C1–C2	DFT	1.439	1.891	−15.5	0.23	−1.755	0.355	1.190
RCP	DFT		0.644	7.8	–	−0.180	1.130	–
5d								
Ni–C_α	DFT	1.959	0.735	4.0	0.21	−0.286	0.769	0.794
	Exp.	1.9543(5)	0.680(9)	7.5(1)	0.78	−0.247	1.137	–
C_α–C_β	DFT	1.497	1.700	−12.7	0.12	−1.417	0.309	1.018
	Exp.	1.4886(7)	1.77(2)	−12.6(1)	0.11	−2.384	0.847	–
C_β–H_β	DFT	1.205	1.387	−11.6	0.07	−1.157	0.251	0.728
	Exp.	1.20(1)	1.33(3)	−5.1(1)	0.15	−1.407	0.789	–
H_β–Ni	DFT	1.653	0.569	6.3	0.96	−0.180	1.089	0.285
	Exp.	1.671(9)	0.553(4)	6.2(1)	1.58	−0.154	1.069	–
RCP	DFT	–	0.507	6.5	–	−0.105	1.098	–
	Exp.	–	0.533	6.3	–	−0.134	1.080	–

[a]The approach developed by Abramov was used to derive $G(\mathbf{r})$ from the experimental $\rho(\mathbf{r})$ distributions [47]

(Fig. 4a, b; Table 4). Note that in the conformer with a staggered Ti–CH_2CH_2–H_β moiety [transition state on the potential energy surface (PES)], the $\angle TiC_\alpha C_\beta$ angle is even smaller (78.6°) than in the eclipsed ground state orientation (84.4°) of the ethyl group despite the lack of any Ti···H_β or Ti···C_β bond paths. This example also reveals the importance of Ti···C_β interaction in β-agostic species, reflected among others by the Ti···C_β delocalization indices which are larger in the staggered version of **6** ($\delta = 0.221$) than in its eclipsed agostic ground state ($\delta = 0.176$).

Accordingly, agostic complexes are typically characterized by a fluxional behavior in solution involving not only internal methyl group rotations (Scheme 4)

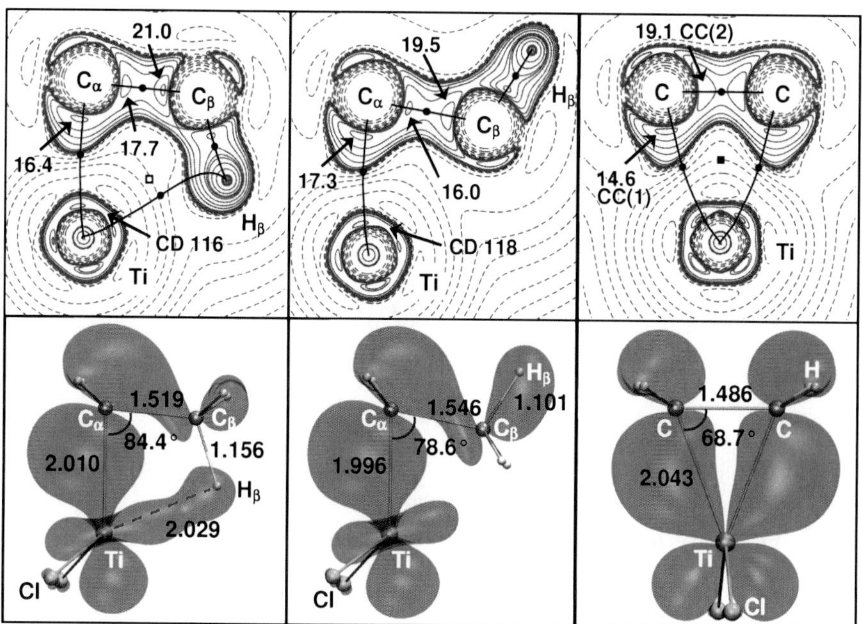

Fig. 4 *Top*: $L(\mathbf{r})$ contour maps in the $TiC_\alpha C_\beta$ plane of (**a**) the eclipsed and (**b**) staggered $[EtTiCl_2]^+$ **6** conformer and of the olefin species $[(\eta^2\text{-}C_2H_4)TiCl_2]$ **7**. Contour lines as specified in Fig. 3; two contour levels deleted (15 and 25 $e\text{Å}^{-5}$) and one extra level at 13 $e\text{Å}^{-5}$ in case of **7**. *Bottom*: isodensity surfaces (0.05 a.u.) of the corresponding HOMOs

Scheme 4 Proposed fluxional behaviour of **5a** in solution

but also olefin hydride elimination steps. The later process is, however, more important for late transition metal alkyls and especially if 4d and 5d metals are involved. Accordingly, in the series $[EtM(d^t bpe)]^+[BF_4]^-$ ($d^t bpe$ = $^tBu_2PCH_2CH_2P^tBu_2$; with M = Ni **5a**, Pd **5a_Pd**, or Pt **5a_Pt**) DFT calculations confirm Spencer's experimental findings [25, 51] that the *cis*-ethene hydride form is only favored in case of the Pt complex of **5a_Pt** (0.84 kJ mol^{-1}) vs. the agostic structure, while the latter one is preferred by the palladium analogue **5a_Pd** by 21.8 kJ mol^{-1}. In case of the nickel complex **5a**, the *cis*-ethene hydride form does not even represent an energetic minimum on the PES but a transition state (61.9 kJ mol^{-1} above the agostic equilibrium geometry) with the olefin moiety aligned perpendicular to the NiP_2H plane [26]. Hence, VT experiments of

$[EtNi(d^tbpe)]^+[BF_4]^-$ can be interpreted in terms of a fast rotation of the β-agostic methyl moiety in solution ($\Delta H^{\ddagger} = 35.1 \pm 1.0$ kJ mol^{-1}, $\Delta S^{\ddagger} = -16 \pm 193$ J mol^{-1} K^{-1}, $E_A = 37.0 \pm 1.0$ kJ mol^{-1}) [26].[5]

However, the C_α–C_β rotational barrier alone cannot be used to derive the agostic stabilization energy. In that case, it is more appropriate to refer to the energy difference, D_{112}, between the agostic ground state geometry and the one defined by an all staggered conformation of the alkyl group showing an MCC angle of 112° [15]. Such a valence angle is a typical value for M–C–C moieties displaying no β-agostic interaction [15]. Typical D_{112} values are in the range of 8–40 kJ mol^{-1} [10, 15] in case of d^0 early transition metal alkyls and are usually significantly larger in case of d^6 or d^8 configurated alkyl complexes (ca. 60 kJ mol^{-1} in case of **5a**). Hence, it is therefore not necessarily appropriate to classify C–H bonds as weak ligands in general [52].

2.1.2 How is the Presence of a M···H Bond Path Influenced by the Nature of the Metal?

In a recent theoretical study using the *extended transition state method* and the *natural orbitals for chemical valance scheme* (ETS–NOCV), Mitoraj et al. suggested that in case of β-agostic interactions the strength of the $M\cdots H_\beta C_\beta$ bond is rather constant in case of cationic Ti(IV)- or Zr(IV)-metallocenes and cationic Ni(II)- and Pd(II)-bis-diimine Brookhart complexes [53]. These theoretical findings are, however, in conflict with experimental results showing that agostic complexes involving late transition metal atoms are at a later stage of the β-elimination pathway than their d^0 early transition metal congeners and are typically characterized by significantly more covalent M···H–C interactions. For example, octahedral d^6 and square-planar d^8 configurated agostic complexes are characterized by the most acute MCC bond angles and highly activated agostic C_β–H_β moieties, while the corresponding main group or d^0 alkyl congeners hardly show any C–H bond activation [4, 10, 26, 52]. This is due to the ease of $M \rightarrow L$ back donation processes involving antibonding π(CC)* and $\sigma(C_\beta H_\beta)^*$ orbitals in late transition metal alkyls (see Sect. 3 for a detailed discussion). Furthermore, also characteristic spectroscopic features like $\nu(C_\beta$–$H_\beta)$ stretching frequencies, ^1H chemical shifts of agostic protons, or $^1J(C_\beta H_\beta)$ coupling constants allow a clear discrimination between agostic d^0-configurated and electron-rich late transition metal complexes. Also the differences in the electronic structures between agostic main group and transition metal complexes should be reflected in the spectroscopic characteristics and charge density distributions of the respective agostic moieties. Indeed, we will show in the following sections that the nature of β-agostic

[5] We note that the barrier of methyl group rotation in the d^8 species **5a** is close to the one computed for our theoretical d^0 model system $EtTiCl_2^+$ (2.8 kJ mol^{-1}) (Fig. 4) suggesting a comparable agostic stabilization in both types of compounds.

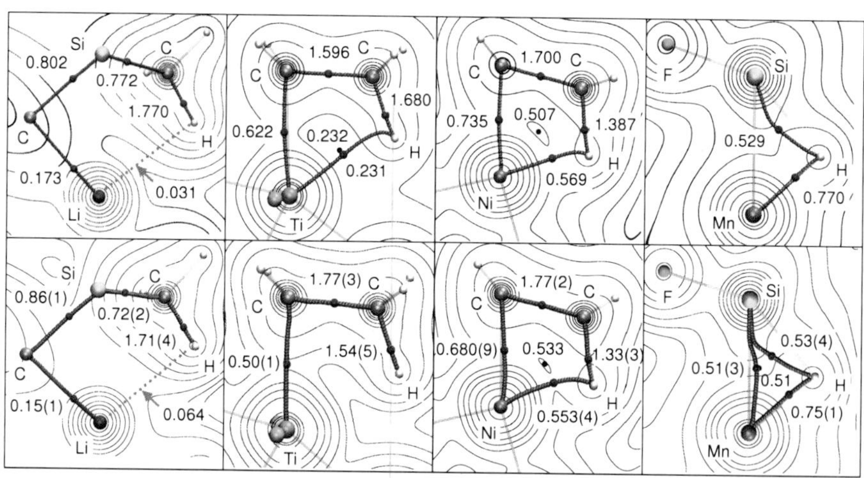

Fig. 5 Ball and stick models of (*top*) the theoretical and (*bottom*) experimental agostic fragments of [2-$(Me_3Si)_2CLiC_5H_4N]_2$ **8** [54, 55], $EtTiCl_3$(dmpe) **2** (experimental model: [11], theoretical model: enforced agostic geometry as shown in Fig. 3), $[(DCpH)Ni(d'bpe)]^+[BF_4]^-$ **5d** [26], and $[Cp'Mn(CO)_2(\eta^2\text{-}HSiFPh_2)]$ **9** [Cp′= $(\eta^5\text{-}C_5H_4Me)$] [49, 56] including the CPs and their corresponding $\rho(\mathbf{r})$-values (in eÅ^{-3}) along the bond paths. The following density contours 2.0×10^n, 4.0×10^n, 8.0×10^n e bohr^{-3} with $n = -3, -2, -1, 0$ were used to illustrate the experimental and theoretical charge density distribution of these benchmarks. The Li-agostic reference system displays no BCP; however, the minimal $\rho(\mathbf{r})$-value and its location (marked by an *arrow*) along the Li···H vector has been specified as reference point. RCPs and BCPs were denoted by *black* and *red spheres*, respectively

interactions yields a continuum between purely closed shell and covalent interactions, and that their strength critically depends on the electronic characteristics of the metal–ligand ML_n moieties involved in the M···$H_\beta C_\beta$ interactions.

Figure 5 shows the experimental and theoretical charge density contour maps of various agostic benchmark systems involving (1) a main group metal (Li^I in **8**), (2) a d^0 configurated transition metal (Ti^{IV} in **2**), (3) a d^8 configurated transition metal (Ni^{II} in **5d**), and (4) a d^6 configurated transition metal (Mn^{II} in **9**) displaying activated η^2-X–H bonds (X = C, Si). Inspection of the coordinating C–H/Si–H bonds in (1)–(4) reveals the expected trend. Accordingly, in case of our main group examples, no BCP can be located between the Li and the coordinating H_γ atoms despite the remarkably short Li···H distance of 2.320(6) Å which represents one of the shortest contacts reported for the so-called lithium agostic compounds [54, 55]. However, the absence of any significant C_γ–H_γ bond activation and lack of significant charge density accumulation in the Li···H_γ coordination region classify this interaction as nearly exclusively electrostatic (see Sect. 4 for details). The situation then changes dramatically if we analyze the electronic situation in the d^0 Ti complex **2**. We noted above that **2** is close to a topological bond catastrophe point. Hence, despite the absence of the Ti···H_β bond path in the relaxed theoretical and experimental multipolar models its presence can be enforced by a subtle shortening of the Ti···H_β distance (by 0.014 Å) in our theoretical model (Fig. 5)

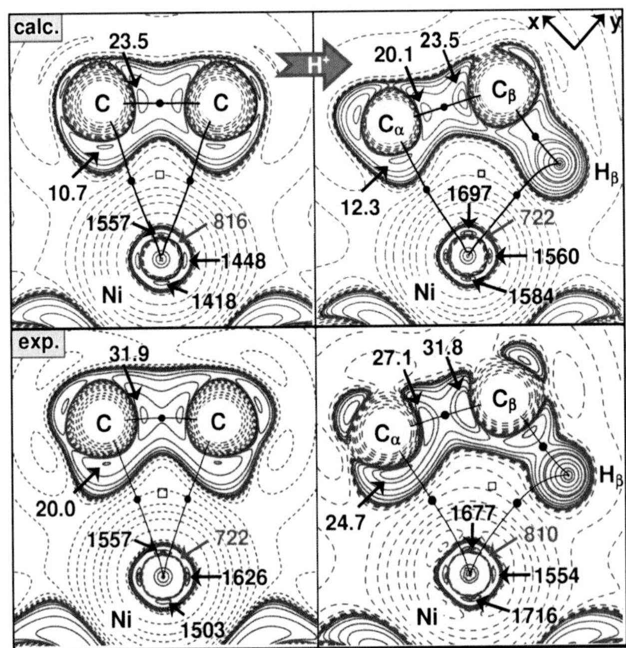

Fig. 6 Experimental contour map of $L(\mathbf{r}) = -\nabla^2\rho(\mathbf{r})$ and bond paths (*black solid lines*) showing (*top*) the change of the calculated $L(\mathbf{r})$ pattern in (DCp)Ni(d′bpe) **4d** upon protonation yielding **5d**; (*bottom*) corresponding experimental $L(\mathbf{r})$ maps of the d^{10} olefin complex (C_2H_4)Ni(d′bpe) **4a** (adopted from [60]) and the protonated agostic d^8 congener $[(DCpH)Ni(d'bpe)]^+[BF_4]^-$ **5d** (adopted from [26]). Contour lines as specified in Fig. 3 but slightly adopted to allow a better visualization of the $L(\mathbf{r})$ fine structure

or via the change of the methodology (e.g., DFT functional) [10, 11]. However, in contrast to our Li-agostic model system, the charge density accumulation is significantly larger in the agostic bonding region of **2** hinting for a more covalent agostic interaction in **2** vs. **8**. Further enhancement of the agostic interaction can be achieved by additional charge polarization as evidenced by our theoretical Ti^{IV} model $EtTiCl_2^+$ **6** leading to the formation of a $Ti^{IV}\cdots H_\beta$ bond path (Fig. 4). However, somewhat surprisingly, despite its short M···H contact (2.029 Å) and significant C–H activation by ca. 0.06 Å; $r(C_\beta{-}H_\beta = 1.156$ Å) the density difference between the RCP of the agostic moiety and the $Ti\cdots H_\beta$ BCP ($\rho(\mathbf{r})_{BCP} = 0.312$ eÅ^{-3}; ($\rho(\mathbf{r})_{RCP} = 0.304$ eÅ^{-3}) in **6** remains small [11, 38] (Table 4). Even in the case of the experimental d^8 nickel complex $[(DCpH)Ni(d'bpe)]^+[BF_4]^-$ (DCp = dicyclopentadiene) **5d** (Fig. 2) – our prototype of a stable agostic complex characterized by a (1) highly activated C–H bond ($\Delta C_\beta{-}H_\beta$ = ca. 0.12 Å relative to a standard aliphatic C–H bond), (2) a remarkably short $Ni\cdots H_\beta$ distance of 1.671 (9) Å, and (3) a highly acute $NiC_\alpha C_\beta$ angle of 74.92(3)° – no significant charge density differences between the Ni···H BCP and RCP of the agostic moiety were found (Fig. 6; Table 5) [26]. Hence, the presence and topological stability of a M···H critical point (CP) seem to be in the best case a rather insensible criterion to classify agostic interactions. However, at least the trend to more covalent M···H bonding in the series **8** < **2** < **6** < **5d** < **9** (Fig. 5) correlates with an increasing charge density accumulation in the metal hydrogen bonding region. Hence, in case of the d^8 nickel species **5d**, we nearly approach the M···H BCP features of the Schubert-type d^6 manganese complex **9** – our experimental reference system for a σ-type complex displaying a nearly fully established hydridic M–H bond path [49, 56]. In the next section, we will outline that the lack of pronounced agostic M···H bond paths in the early and late transition metal alkyls is a natural consequence of the delocalized character of the agostic interaction and not a failure of Bader's bond path concept.

3 The Agostic Interaction in the MO Picture

To understand the reasons for the elusive nature of β-agostic M···H_β bond paths, we analyzed the salient molecular orbitals of the simplest reference systems of agostic early and late transition metal alkyls: the cations $[EtCa]^+$ **10** (d^0) and $[EtNi]^+$ **11** (d^8) (Fig. 7). We first note that only $EtNi^+$ is characterized by the presence of more than one molecular orbital contributing significantly to metal–ligand interactions. In case of $[EtCa]^+$, HOMO-2 and HOMO-3 represent nearly exclusively ligand orbitals characterized by vanishing coefficients at the metal atom. Another difference is evident from inspection of the energetically highest-lying molecular orbitals involved in the agostic interaction: the HOMO in $[EtCa]^+$ and HOMO-4 in case of $[EtNi]^+$. The HOMO of $[EtCa]^+$ may be described as a Ca–C_α σ-bonding orbital formed by combination of the metal-based $3d_z{}^2$ atomic orbital (with significant admixture of the 4s orbital) with the C–H bonding but C–C antibonding $\pi_{z'}$ orbital of the ethyl fragment. We note that the ethyl group in $[EtCa]^+$ has been canted in such a way that there is now a positive overlap between the ring- or doughnut-shaped density contours of the Ca ($4s/3d_z{}^2$) hybrid orbital and the $C_\beta H_\beta$ fragment establishing Ca···H_β bonding [11]. Accordingly, the HOMO of $[EtCa]^+$ represents the delocalization of the Ca–C_α bonding electron pair over the metal alkyl backbone, while all other valence orbitals represent (distorted) σ/π bonding orbitals of the ethyl ligand.

The agostic interaction in early transition metal complexes can therefore be regarded as a *covalent interaction which can be understood only in terms of a M–C_α bonding orbital that is delocalized over the entire ethyl group: the reduction of the MCC valence angle allows the metal atom to establish bonding interactions with the β-C and its appended H atom. The geometry of the M–Et interaction is such that bonding to both C_α and C_β is effected by the same orbital on M* [15].

More precisely, the reduced antibonding character of the C–C antibonding $\pi_{z'}$ orbital relative to the $[C_2H_5]^-$ **12** anion among other observations (see Sects. 4 and 5) led SMG to conclude that agostic stabilization in early d^0 transition metal alkyls *arises from negative hyperconjugative delocalization of the M–C bonding electrons. Accordingly, the bonding between the metal atom and the ethyl group in d^0 transition metal alkyls is effectively established by one electron pair/molecular orbital* [10].

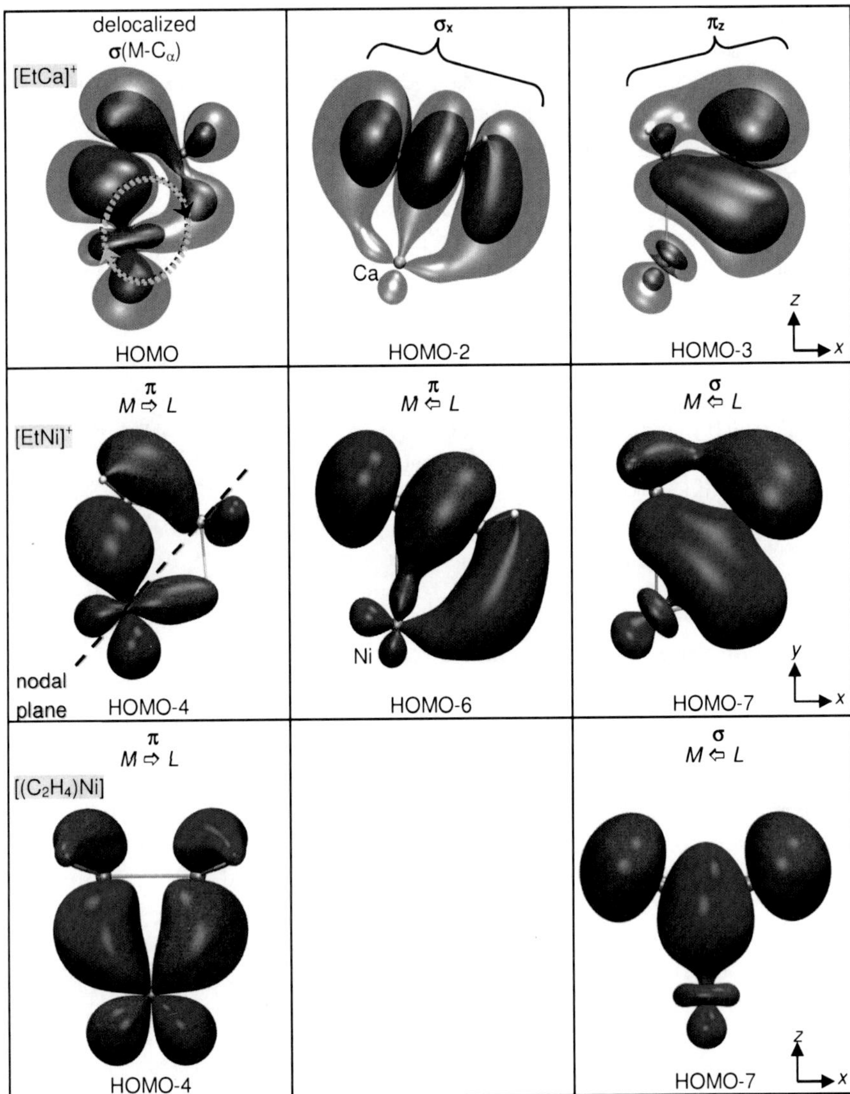

Fig. 7 Salient molecular orbitals in $[EtCa]^+$ **10** and $[EtNi]^+$ **11** and the olefin complex $[\eta^2\text{-}(C_2H_4)Ni]$. The default isovalues (depicted as opaque isosurfaces) were drawn at 0.05 a.u. to allow a direct comparison of the electronic structures displayed by the three systems. *Top*: the HOMO in $[EtCa]^+$ exhibits the delocalization of the σ(Ca–C_α) electron pair over the metal alkyl backbone, while all its other valence orbitals represent mainly (distorted) σ/π bonding orbitals of the ethyl ligand. Only at lower isodensity values (transparent envelopes), admixtures of Ca(4s/3d) orbitals become noticeable and reveal the resemblance of the electronic structures of early and late agostic transition metal complexes. *Middle*: multicenter molecular orbitals in the $[EtNi]^+$ cation establishing the M $\rightarrow$ L π back donation, M $\leftarrow$ L π donation, and M $\leftarrow$ L σ donation, respectively (L = alkyl unit). *Bottom*: σ donor and π back donation component of the classical DCD model in the $[\eta^2\text{-}(C_2H_4)Ni]$ **13** olefin complex

In the case of our d^8 complex, however, the corresponding orbital (HOMO-4) shows a related but slightly different character. It is characterized by a nodal plane which is oriented perpendicular to the molecular plane and parallel to the $Ni\cdots C_\beta$ vector. Due to these symmetry restraints imposed, HOMO-4 is still contributing simultaneously to *covalent* $Ni–C_\alpha$ and $Ni–H_\beta$ bonding but displays a $C_\beta–H_\beta$ antibonding character. Accordingly, HOMO-4 in $[EtNi]^+$ is mainly responsible for the $Ni \rightarrow L$ π back donation and less supportive for negative hyperconjugative delocalization. As a consequence of the $C_\beta–H_\beta$ antibonding character of HOMO-4, the $Ni \rightarrow L$ back donation process enhances the activation of the β-agostic C–H bond relative to the d^0 type congeners. This conforms with the observation that β-agostic late transition metal alkyls usually display larger activated $C_\beta–H_\beta$ bonds and more covalent $M\cdots H_\beta$ interactions relative to their d^0 congeners. Therefore, the latter ones typically rest at the early stage of the β-H elimination pathway due to their reduced π back donation capabilities. Figure 7 also reveals that the agostic interaction in $[EtNi]^+$ is further supported by additional $M \leftarrow L$ π donation (HOMO-6) and $M \leftarrow L$ σ donation (HOMO-7). Accordingly, the bonding in β-agostic transition metal complexes can be described by a generalized Dewar–Chatt–Duncanson (DCD) model [57, 58]. In that case the additional $M \leftarrow L$ π donation (which complements the $M \leftarrow L$ σ-donation component of the classical DCD model for olefin complexes) simply reflects the increased functionality of β-agostic alkyl ligands by establishing additional $M\cdots H_\beta$ interactions.[6]

If we now relate the MO picture with our charge density observation, we can identify the $M \leftarrow L$ σ donation as a main source of the charge density accumulation inside the agostic moiety, especially in case of electron-rich agostic transition metal complexes. Hence, the small differences in charge accumulation between the RCP and the $M\cdots H$ BCP (which might even result in a bond catastrophe scenario; see Sect. 2.1) are just a natural consequence of the competing *$M \leftarrow L$ donation and $M \rightarrow L$ back donation* processes in agostic moieties. It is therefore not advisable to correlate the strength of the agostic interaction only with the magnitude and presence of the $M\cdots H$ BCP. *The topological instability of agostic $M–H_\beta$ bond paths simply reflects the delocalized nature of agostic bonding*. In this respect, the bond catastrophe scenario in agostic interactions is related to the topological peculiarities displayed by transition metal π-complexes. Here, the stability of a cyclic ring vs. a T-shaped structure of the bond paths critically depends on relative magnitude of both DCD components in the MCC units ($M \leftarrow L$ σ donation vs. $M \rightarrow L$ π back donation; L = π-ligand) [59–63]. Also in these cases the topology of the bond path (T-shaped vs. cyclic) alone cannot be taken as a sole criterion to evaluate the strength of the delocalized interaction between the metal and its

[6] On the basis of this concept, β-agostic interactions can be clearly discriminated from *σ-complexes* formed by metal centers and η^2-coordinating X–H moieties (e.g., X = H, B, C, Si). In the latter case, only bonding and antibonding σ(X–H) orbitals are involved in the metal interaction, while β-agostic compounds are characterized by the additional delocalization of the $M–C_\alpha$ bonding pair via negative hyperconjugation.

π-ligand [64–66]. As another consequence, the true hapticity of metal π-complexes is usually not reflected in the bond path topology as shown by Farrugia et al. and warrants more sophisticated charge density analyses (e.g., inspection of the virial graph) [67].

4 Control Parameters (I): Negative Hyperconjugation

In this section, we will outline (1) the driving forces for the negative hyperconjugative delocalization process in β-agostic complexes, (2) its control parameters, and (3) the charge density criteria to quantify the extent of electron delocalization in various benchmark systems.

4.1 The Driving Force of Negative Hyperconjugation in β-Agostic Compounds

Negative (anionic) hyperconjugation (NHC) is often referred to as the generalized anomeric effect [68, 69] and was originally proposed by Roberts [70] to describe the electronic effects of $\pi \rightarrow \sigma^*$ delocalization. The electronic consequences of negative hyperconjugation can be demonstrated in case of the non-agostic reference system EtLi **14** and $[EtCa]^+$ **10**, which displays all characteristics of a β-agostic compound. Inspection of the HOMOs of ($[CH_2CH_3]^-$ **12**, **14,** and **10** (Fig. 8) reveals that the C–C antibonding character becomes significantly reduced in comparison with the HOMO of ethane (π^* orbital) due to negative hyperconjugation (Scheme 5).

Although the C–H bonds of CH_3 groups are usually considered to be σ systems, they possess orbitals of π symmetry which can interact with the $p_\pi(C_\alpha)$ on adjacent atoms [71]. Accordingly, each $p_\pi(C_\alpha)$ orbital will delocalize into a linear combination of the $\sigma^*(CH)$ antibonds of π symmetry in accordance with a negative hyperconjugative $\pi \rightarrow \sigma^*$ delocalization (see for example [72]). In carbanionic systems, negative hyperconjugation thus involves the interactions of the occupied lone pair orbital $p_\pi(C_\alpha)$ at the carbanion with the occupied $\pi(C_\beta H_\beta)$ and the vacant $\pi^*(C_\beta H_\beta)$ orbitals (Scheme 5) [55, 71]. As a consequence the reduced π^*(C–C) antibonding character is clearly signaled by the diminished LCAO coefficient in the HOMO at the C_β atom (Scheme 5; Fig. 8). Even though the $p_\pi(C_\alpha)$–$\pi^*(C_\beta H_\beta)$ interaction is weak in case of the rather electropositive H_β atom, it causes a significant $\pi \rightarrow \sigma^*$ delocalization and reduction of the carbanionic character at the C_α atom (see below). However, the charge transfer from the C_α carbanion to the H_β atom is small causing only a minute C_β–H_β bond activation in EtLi in comparison with ethane (1.108 Å vs. 1.099 Å). In organometallic chemistry, it is, however, a well-documented phenomenon that replacement of the β-hydrogen atom by a

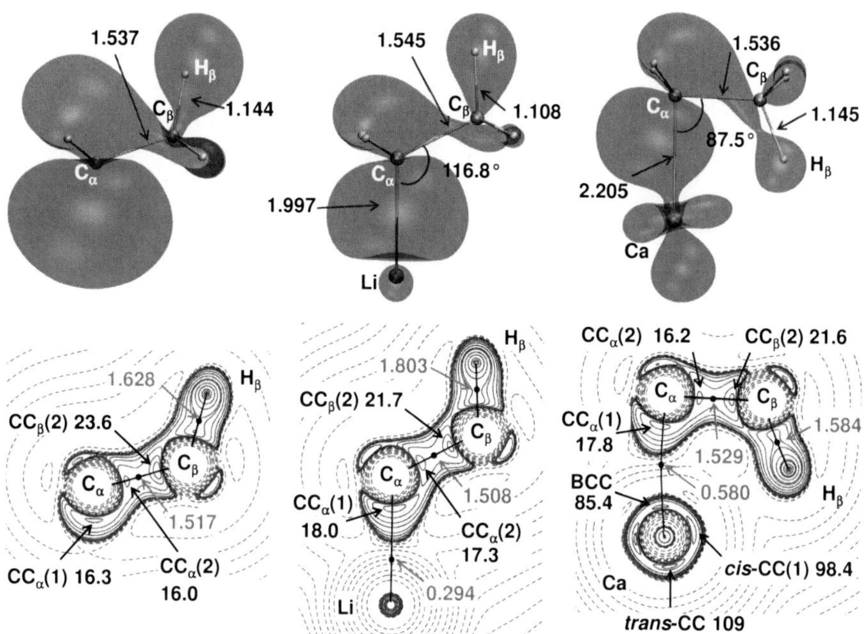

Fig. 8 *Top*: highest occupied molecular orbital (HOMOs) of the ethyl carbanion $[Et]^-$ **12**, EtLi **14**, and the $[EtCa]^+$ **10** cation (default isovalues at 0.05 a.u.) characterized by a reduced π^*(C–C) antibonding character due to negative hyperconjugation. Bond lengths are specified in Å. *Bottom*: theoretical $L(\mathbf{r}) = -\nabla^2\rho(\mathbf{r})$ contour maps (eÅ^{-5}), bond paths (*black solid line*), and selected values: *gray* at the CPs (eÅ^{-3}), *black* at the CCs (eÅ^{-5}) in the molecular $C_\alpha C_\beta H_\beta$ plane. Default contour lines and notations as specified in Fig. 3

more electronegative atom (e.g., a fluorine ligand) causes a more pronounced charge transfer and in the extreme case of $[CH_2CH_2F]^-$ (Scheme 6) even the rupture of the C_β–F bond [71].

As a consequence of orbital interactions outlined in Scheme 5, the extent of negative hyperconjugation will be optimized at $\tau(MC_\alpha C_\beta H_\beta) = 0°$ or 180°, falling to zero at $\tau = 90°$. This elegantly explains the observation that the C–H_β activation in agostic compounds is only pronounced for β-hydrogen atoms located inside the $MC_\alpha C_\beta$ molecular plane.

A related situation occurs in case of β-agostic d^0 transition metal alkyls. Here, the negative hyperconjugative delocalization of the M–C_α electron pair over the metal alkyl backbone (Fig. 7) is driven by the carbanionic character of the alkyl ligands. However, in contrast to organometallic lithium or other main group complexes, transition metal alkyls use s/d hybrid orbitals at the metal side to furnish a σ-bond with the $p_\pi(C_\alpha)$ lone pair orbital. Due to the nodal structure of the d-type orbitals, the canting of the ethyl group can now occur in such a way that there is a positive overlap between the ring- or doughnut-shaped density contours of the M ($4s/3d_z{}^2$) hybrid orbital and the $C_\beta H_\beta$ fragment (Fig. 7 and 8). Hence, the d-orbital involved in the σ(M–C_α) bond establishes in parallel also the covalent M···H_β

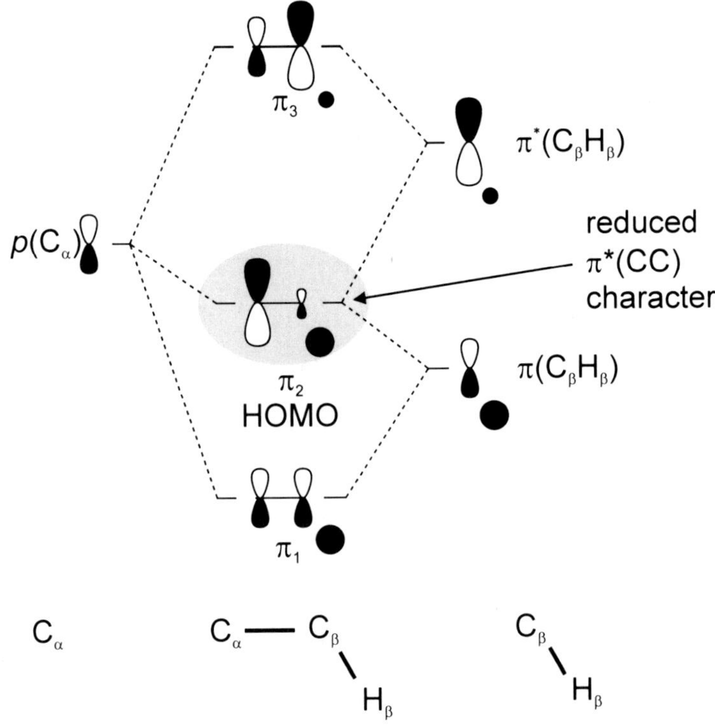

Scheme 5

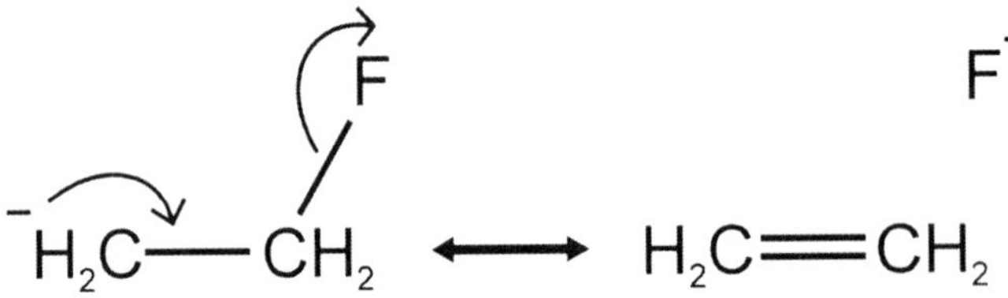

$^{-}H_2C\text{—}CH_2(F) \longleftrightarrow H_2C{=}CH_2 \quad F^{-}$

Scheme 6

bonding [11]. Hence, *the additional $M\cdots H_\beta$ interaction triggers the extent of negative hyperconjugation in a related way as the replacement of H_β by a more electronegative ligand.*

To summarize: the agostic deformation of the alkyl backbone becomes initiated by the action of negative hyperconjugation. This delocalization process of the $\sigma(M–C_\alpha)$ electron pair over the agostic entity is driven by the motivation of the system to reduce its carbanionic character. However, due to the lack of d-acceptor orbitals, main group alkyls cannot efficiently stabilize β-agostic interactions. As a consequence, significant agostic interactions are usually not observed in main group alkyls. We also note at this stage that hyperconjugative delocalization becomes less

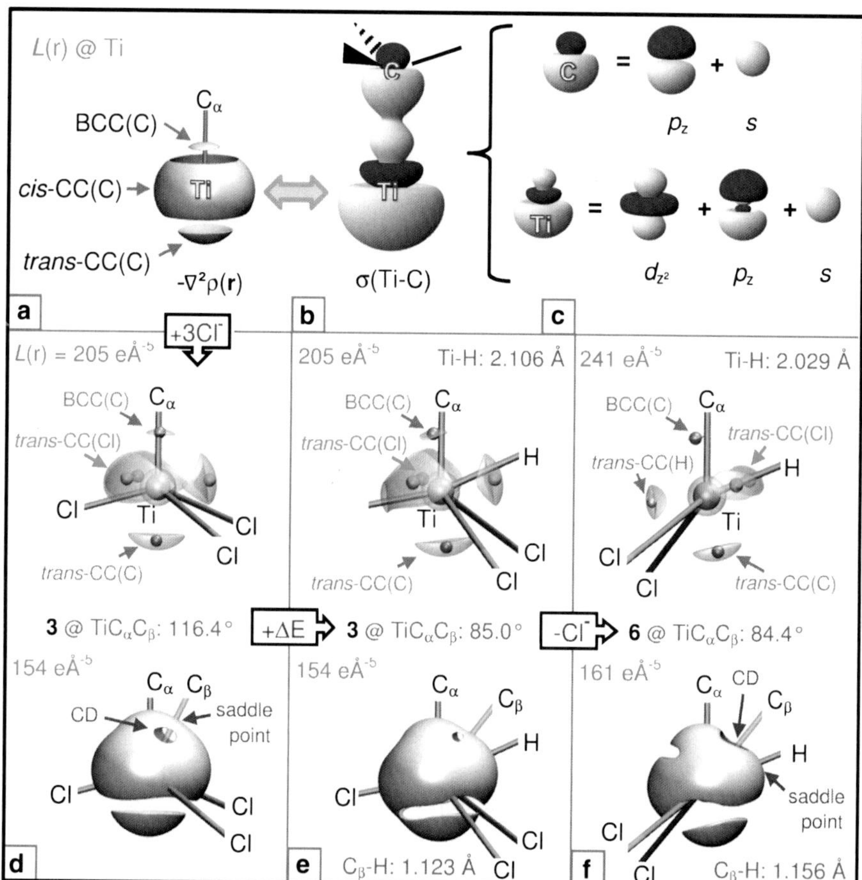

Fig. 9 (**a**) Schematic illustration of the origin of the ligand-induced charge concentrations [BCC (C), *trans*-CC(C), and *cis*-CC(C)] at the Ti atom in $EtTiCl_3$ **3**. These charge concentrations are an integral part of the σ(Ti–C) bond and reflect the nature of metal orbitals involved (19% s-, 7% p-, and 74% d-character) in the metal to ligand bonding. The BCC is less pronounced than its ligand-opposed *trans*-CC congener due to the polarization of the σ(Ti–C) bond; (**b**) schematic drawing of the polarized σ(Ti–C) NBOs; (**c**) schematic analysis of the origin of the σ(Ti–C) bond polarization by the formal hybridization of the metal and C_α atoms; (**d**) $L(\mathbf{r})$ iso-value maps of **3** (based on the total charge density distribution of **3**) in the valence region of the Ti atom (non-agostic equilibrium geometry); (**e, f**) corresponding $L(\mathbf{r})$ maps of **3** in a constrained agostic geometry ($\angle$TiCC = 85°; eclipsed ethyl group orientation) and of $[EtTiCl_2]^+$ **6**. Note that only in case of **6** which displays a β-agostic equilibrium structure a local Lewis-acidic center (denoted "CD") is facing the coordinating C_β-atom. $L(\mathbf{r})$ contour values are specified in (eÅ^{-5}); the coordination vectors to the ligand atoms are only indicated and are not true to scale

important as a driving force of agostic interactions in case of electron-rich late transition metal complexes, where $M \rightarrow L$ π back donation processes trigger the establishment of covalent M···H–C interactions (see Sect. 5). Hence, negative hyperconjugative delocalization is an important control parameter of agostic interactions especially in case of main group and d^0 transition metal complexes.

4.2 *How to Quantify Hyperconjugation by Charge Density Analyses*

The extent of negative hyperconjugation should be correlated with the diminished magnitude of the valence shell charge concentration (VSCC) in the lone pair region at the carbanionic C_α atom [11, 55]. Indeed, analysis of the negative Laplacian, $L(\mathbf{r}) = -\nabla^2\rho(\mathbf{r})$ of theoretical charge density distributions shows the expected decrease of the corresponding VSCC [denoted $CC_\alpha(1)$ in Fig. 8] in a series of alkyl complexes. All values are specified in (eÅ^{-5}) and **3_ago** refers to the nonequilibrium charge density distribution of $EtTiCl_3$ calculated by imposing a constrained agostic conformation with $\angle$TiCC = 85° (Fig. 9e). **3_eq** refers to the non-agostic equilibrium geometry of $EtTiCl_3$ as depicted in Fig. 1.

3_eq (18.7) > **3_ago** (17.2) > **2** (15.7) > **7** (14.6) ≫ **5d** (12.3) > **4d** (10.7)

Accordingly, the depletion of $CC_\alpha(1)$ in the lone pair region of the carbanion indicates a redistribution of charge within the valence shell of C_α. This observation can therefore be used as a sensitive measure of the extent of electron delocalization over the alkyl backbone by negative hyperconjugation [11, 55]. Hence, enforcing an agostic conformation in $EtTiCl_3$ leads to an enhancement of the NHC and a significant reduction in the charge concentration at $CC_\alpha(1)$. In the dmpe complex displaying a less polar Ti–C bond, the $CC_\alpha(1)$ becomes further depleted (Fig. 3). The lowest $CC_\alpha(1)$ values are consequently observed in the titanium olefin complex **7** and the nickel complexes **5d** and **4d** characterized by an even more covalent M–C bonding (Figs. 4 and 6). We note that also the agostic cations $[EtCa]^+$ ($CC_\alpha(1)$ = 17.8 eÅ^{-5}) and $[EtTiCl_2]^+$ ($CC_\alpha(1)$ = 16.4 eÅ^{-5}) obey this trend – despite the partial hindrance of M–C delocalization by the additional charge polarization of the M–C bond (Figs. 4 and 8). It is also interesting to note that the staggered version of $[EtTiCl_2]^+$ displays a larger $CC_\alpha(1)$ value (17.3 eÅ^{-5}) than its β-agostic counterpart – despite its more acute $\angle$TiCC angle of 78.6° vs. 84.4°, respectively (Fig. 4). This clearly underpins the important role of Ti···H interaction to enhance the NHC effect as dominating driving force of agostic interactions in early transition metal complexes. Hence, the extent of β-agostic interactions and of negative hyperconjugation can be correlated even in systems with highly polar M–C_α bonds.[7]

[7] The hindrance of NHC delocalization by charge polarization is documented by a comparison of the situation in $[CH_2CH_3]^-$ **12** and $[CH_2CH_3]^-[Li^+]$ **14**. The large NHC in **12** vs. **14** is reflected not only by a smaller $CC_\alpha(1)$ charge concentration of 16.3 eÅ^{-5} in **12** but also by a remarkable activation of the C_β–H_β bond *trans* to the C_α lone pair (ca. 0.04 Å) in **12** which is less pronounced in **14** (ca. 0.01 Å). As another consequence, $CC_\alpha(1)$ in the agostic $[EtCa]^+$ cation is only slightly

Another criterion arises from the fact that the asymmetry of the charge density distribution along the C_α–C_β bond path becomes more and more balanced with increasing olefinic character of the ethyl moiety. This again can be measured by determining the relative magnitudes of the bond-directed charge concentrations $CC_\alpha(2)$ and $CC_\beta(2)$ at the C_α and C_β atom, respectively. An increasing olefinic character is then signaled by a reduction in the differences between the magnitudes of $CC_\alpha(2)$ and $CC_\beta(2)$. Accordingly, the pronounced carbanionic character of EtLi yields a rather large difference in the relative magnitudes of $CC_\alpha(2)$ (17.3 eÅ^{-5}) and $CC_\beta(2)$ (21.7 eÅ^{-5}) (Fig. 8). This difference becomes smaller in the agostic dmpe complex **2** with $CC_\alpha(2) = 18.5$ eÅ^{-5} and $CC_\beta(2) = 21.3$ eÅ^{-5} (Fig. 3) yielding values close to the balanced CC(2) values (19.1 eÅ^{-5}) displayed by the Ti^{IV} olefin complex η^2-(CH_2CH_2)$TiCl_2$ **7** (Fig. 4) [11]. Accordingly, the magnitudes of $CC_\alpha(1)$, $CC_\alpha(2)$, and $CC_\beta(2)$ provide another sensitive charge density measure to evaluate how far an agostic system has propagated along the β-H elimination reaction coordinate.

Pioneering studies by Bader, Cremer, and coworkers related conjugative interactions also to the presence of bond ellipticity (ε) – another charge density-based criterion [73, 74]. According to these studies, C–C bonds affected by hyperconjugation might be identified by bond orders $n > 1$ and bond ellipticities $\varepsilon > 0$ at the respective BCPs [74]. However, SMG showed that only the evaluation of complete ellipticity profiles $\varepsilon(\mathbf{r})$ along the C_α–C_β bond paths provides the salient information to quantify the extent of hyperconjugative electron delocalization [54, 55]. Especially the carbanionic character of the C_α atoms and the hypercoordinative nature of the C_β atoms displaying short $M\cdots C_\beta$ contacts are not fully reflected in the ellipticity values at the C_α–C_β BCP alone. Figure 10 shows the ellipticity profiles of various model systems. According to the mathematical definition (inset of Fig. 10), ε values greater than zero indicate partial π-character (or other electronic distortion away from σ-symmetry) along the bond path. This is for example revealed by the characteristic ellipticity profile of the benchmark system ethene **16** along the C_α–C_β bond path (Fig. 10).

We note that C_2H_4 shows a bell-shaped ellipticity profile around the BCP characteristic of a C=C double bond, whereas C_2H_6 **15** exhibits zero ellipticity along the whole bond path, indicating no deviation of $\rho(\mathbf{r})$ from σ-symmetry. However, the situation is already more complex for the Ti^{IV} olefin complex [η^2-(CH_2CH_2)$TiCl_2$] **7** showing *two* instead of *one* maxima as in the case of C_2H_4 **16**. This puzzling feature is, however, related to the presence of the Ti–C bond-directed charge concentrations [CC(1) in Fig. 4] in the MCC plane. These two Ti–C directed charge concentrations at the carbon atoms (which are lacking in **16**) simply result from the charge density redistribution in the valence shell of both carbon atoms – formally an sp^2/sp^3 rehybridization – due to the weak but noticeable Ti $\rightarrow \pi^*$(CC) back donation in **7**. Accordingly, the presence of two ellipticity maxima along the C_α–C_β bond path in an agostic moiety is interwoven with the presence of NHC

smaller (17.8 eÅ^{-5}) compared with the one in the neutral non-agostic lithium congener (18.0 eÅ^{-5}) (Fig. 8).

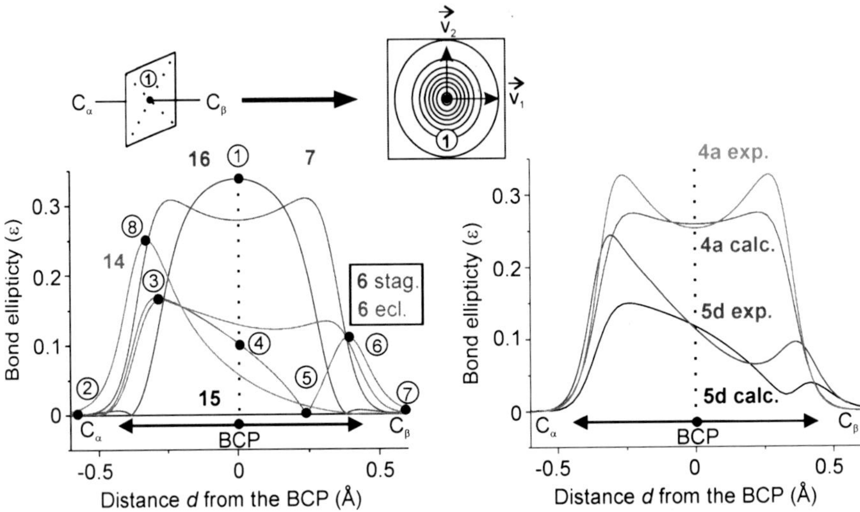

Fig. 10 *Left*: calculated bond ellipticity profiles along the C_α–C_β bond path of $[EtTiCl_2]^+$ (**6**; eclipsed and staggered conformation) in comparison with C_2H_6 **15**, C_2H_4 **16**, EtLi **14**, and $[(\eta^2\text{-}C_2H_4)TiCl_2]$ **7** (adopted from [10]). The definition of the bond ellipticity is illustrated by the $\rho(\mathbf{r})$ contour map in the *upper right corner*, showing the charge density in the plane perpendicular to the bond path at the C–C bond CP of C_2H_4 (denoted "①" in this figure). $\varepsilon(\mathbf{r})$ values larger than zero are thus a measure of the deviation from a cylindrical charge distribution $\rho(\mathbf{r})$ along a bond path: $\varepsilon = \lambda_1/\lambda_2 - 1$ (with $\lambda_1 < \lambda_2 < 0$). λ_i are eigenvalues of the corresponding eigenvectors $\mathbf{v}_1$ and $\mathbf{v}_2$ of the Hessian matrix of $\rho(\mathbf{r})$; see [11] for further details. *Right*: experimental and calculated ε profiles along the C_α–C_β bond path of the olefin complex $(\eta^2\text{-}C_2H_4)$Ni(d^{t}bpe) **4a** and the agostic alkyl spezies $[(DCpH)Ni(d^tbpe)]^+[BF_4]^-$ **5d**

delocalization which renders an alkyl complex in the extreme case into an olefin hydride species (see [11] for a detailed discussion). Hence, the appearance of $\varepsilon(\mathbf{r})$ maxima in the proximity of the C_β atom directly signals the establishment of $M\cdots C_\beta$ interactions as a consequence of the β-agostic delocalization process. Indeed, inspection of the $\varepsilon(\mathbf{r})$ profiles of our agostic d^0/d^8 model systems **6** and **5d** reveal pronounced $M\cdots C_\beta$ interaction which are lacking in the non-agostic lithium alkyl **14** (Fig. 10) [11, 55]. Furthermore, analysis of the ellipticity profiles allows a clear discrimination of electronic situation in the β-agostic cations **6_ecl** and its staggered variant **6_stag** (Figs. 4 and 10). The close relationship of the ellipticity profiles of **7** and **6_stag** shows that the olefinic character in **6_stag** is already well established in accordance with its highly acute ∠TiCC angle of 78.6°. The same is true for the late agostic reference system **5d** while the ε-profile of the β-agostic complex **6_ecl** rather conforms to an electronic situation at the early stage of the β-H elimination process.

Similar conclusions were reached in case of the d^2 niobium complex Tp^{Me2}NbCl (MeCCMe)(^{i}Pr) by Pantazis et al. where *delocalization of the electrons in the Nb–C_α bond into a π^* orbital of the alkyne ligand causes the Nb–C_α σ-bonding orbital to tilt of the Nb–C axis* thereby initiating the canting at the α-carbon atoms and the establishment of α- or β-agostic interactions [41]. Hence, reduction of the carbanionic character at the C_α atom via delocalization of the M–C electron pair at the C_α atom seems to be one common driving force for α- and β-agostic interactions despite the assumed lack of direct C–H donation to the metal in the former case [75, 76]. We finally note that the bond ellipticity profile criterion has been successfully used in the mean time to study also the perturbance of C–C bonds by so-called M ← σ(C–C) agostic interactions in an experimental charge density study of the titanacyclobutane complex $Ti(C_5H_4Me)_2[(CH_2)_2CMe_2]$ [9]. For detailed experimental studies on the distribution of charge over the anionic ligand backbone and characterization of the carbanionic character in highly polar Li^+R^- type complexes, see [77–80]. Experimental and theoretical ellipticity profiles were also used to study the electron π-delocalization in acyclic and *N*-heterocyclic carbenes and their corresponding metal complexes [81, 82]. Hence, bond path profiles provide versatile information in general to study electron (de)localization phenomena and/or the perturbance of the charge density in σ-bonds by polarization effects in the proximity of VSCCs [83].

4.3 Control of the Extent of Negative Hyperconjugation and Agostic Interactions by the Nature of the Ligands

As outlined above, the nature of metal (main group metal vs. electron-poor or electron-rich transition metal) is already one important control parameter of the extent of negative hyperconjugation and thus of the nature of agostic bonding. Other control factors are provided by the substitution of the C_α and C_β atoms by other elements.

4.3.1 β-Agostic Alkyl and Amido Complexes

In this section we will investigate how a systematic hindrance of hyperconjugative delocalization in the ligand backbone reduces the extent and nature of the β-agostic interaction. Hindrance of the delocalization process is efficiently accomplished in case of amido complexes where the metal ligand bonding is stabilized rather by direct M ← N π-donation than via negative hyperconjugation. As a consequence, the electronic motivation to establish β-agostic interactions or the disposition for β-H elimination is clearly reduced in amido complexes. Accordingly, β-H elimination represents a rarely documented phenomenon for transition metal amido complexes [84–86]. The first directly observable example of a β-H elimination involving the monomeric late transition metal amido complex, $[Ir(PPh_3)_2(CO)(N(CH_2Ph)Ph)]$, was reported by Hartwig in 1996 [84] yielding the stable *N*-phenyltoluenimine and the hydrido species $[Ir(PPh_3)_3(CO)H]$. Hartwig suggested that *β-H elimination of late metal amides can be much slower than elimination of the*

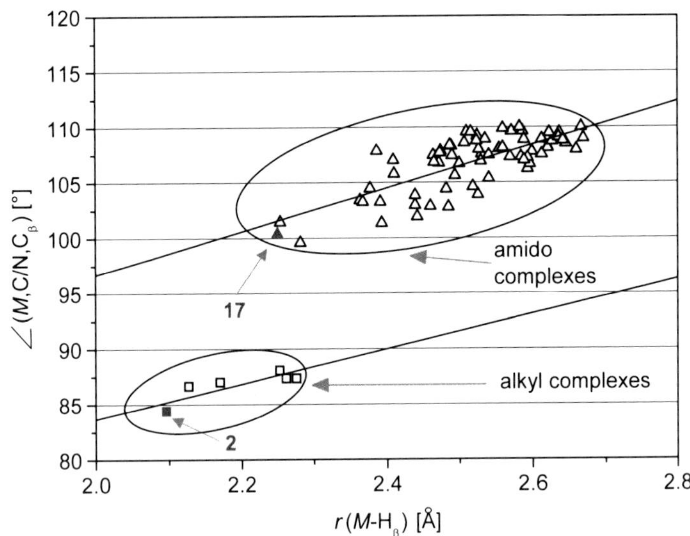

Fig. 11 Plot of the $\angle$(M, C/N, C_β) angles vs. the r(M–H_β) distances for potential d^0 β-agostic alkyl and amido complexes surveyed in a CSD search (excluding disordered structures and data yielding $R1$-values <7.5%). The data point for our amido and alkyl agostic benchmarks CpTiN$(CHMe_2)_2Cl_2$ **17** [89, 90] and $ETiCl_3$(dmpe) **2** [11], respectively, are marked by *red symbols*

corresponding alkyl complexes [84]. Accordingly, *the different mechanistic features detected for amido, alkoxo, and alkyl derivatives which make it difficult to directly compare their tendencies to undergo β-elimination reaction* [87] are, however, already reflected in their ground state geometries. Indeed, a search in the Cambridge Structural Database (CSD) [88] reveals that agostic alkyl complexes are characterized by significantly shorter M···H contacts and smaller ∠MCC angles in comparison with agostic amido complexes (Fig. 11).

Detailed inspection of the β-agostic moiety in CpTiN$(CHMe_2)_2Cl_2$ **17** (Figs. 11 and 12) shows that the N_α–C_β bond of the agostic N-$(CHMe)_2$ moiety is actually larger in comparison with its non-agostic entity (Fig. 12). This observation suggests that negative hyperconjugative delocalization – which leads as in the case of our β-agostic alkyl benchmark, $ETiCl_3$(dmpe), to a shortening of the C_α–C_β bond – might not play any important role in amido complexes. A literature survey further reveals that the observed lengthening of the N_α–C_β bond is a characteristic feature of agostic amido moieties which thwarts the β-H elimination process by hindering the amido ligand from attaining an imine-type character [90]. Obviously, the pronounced π-bonding character of the coordinating N atom hinders hyperconjugative electron delocalization of the M–N bonding electrons which is essential for establishing strong β-agostic interactions. Similar conclusion was reached by Pillet and coworkers in an experimental charge density study of the zirconocene Zr(2, 4-C_7H_{11})[(*i*-Pr)NCHPhCH_2CMe=CH_2] (C_7H_{11} = dimethylpentadienyl) **17a** [91]. The Zr(*i*-Pr) moieties displays similar structural parameters (∠(ZrNC_β = 102.87 (4)°; N–C_β = 1.4710(8) Å, Zr···H_β = 2.4604(6) Å) as the agostic Ti(*i*-Pr) unit in **17** and a vanishing ellipticity of 0.01 at the N–C_β BCP ruling out any significant

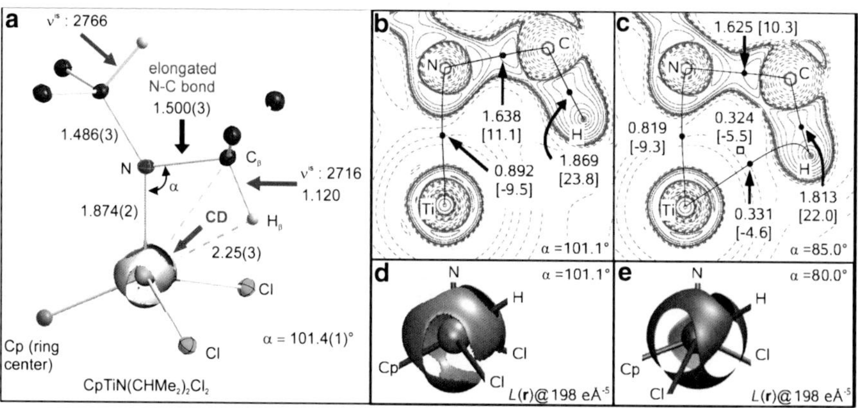

Fig. 12 (**a**) ORTEP representation (50% probability level) of the agostic moiety of CpTiN $(CHMe_2)_2Cl_2$ **17** at 9 K. Salient bond distances (Å) and angles (°) are given together with the experimental value of the isolated C–H stretching frequency, ν^{is} (cm^{-1}). Theoretical isosurface *L* (**r**) map @ 198 eÅ^{-5}, showing a clearly reduced charge concentration in the valence shell of the metal atom opposite the agostic C_β–H_β entity; (**b**, **c**) *L*(**r**) contour maps and bond paths (*black solid line*) in the Ti–N–C plane displaying a $\angle TiNC_\beta$ angle of $\alpha = 101.1°$ and 85°, respectively. Default contour lines and notation as specified in Fig. 3 (one extra level at 191 eÅ^{-5}); (**d**, **e**) corresponding *L*(**r**) envelope maps for the optimized and restrained geometry of **17**, respectively

imine character. Hence, the electronic structures (and the driving forces) of agostic amido and alkyl complexes differ clearly and should be discriminated from one another.

Topological analyses of the theoretical charge density distributions in CpTiN $(CHMe_2)_2Cl_2$ and related model systems show that they all lack the presence of a M···H BP [90] in agreement with the experimental study on **17a** [91]. Only by enforcing shorter M···H contacts via imposition of acute $\angle MNC_\beta$ valence angles, the formation of a Ti···H bond path could be established in the calculations [90] (Fig. 12). Indeed, the enforced M···H BCP in $CpTiN(CHMe_2)_2Cl_2$ shows all the characteristic features of agostic M···H interactions in early transition metal complexes (Table 4). However, the experimental infrared data for the equilibrium structure of **17** reveals two isolated stretching frequencies, ν^{is} [16], for the agostic ($\nu^{is} = 2{,}716\ cm^{-1}$) and non-agostic ($\nu^{is} = 2{,}766\ cm^{-1}$) methine groups suggesting only a marginal bond length difference of 0.005 Å between the agostic and non-agostic C_β–H bonds (1.120 and 1.115 Å, respectively). This is in line with the corresponding $^1J_{CH}$ coupling constants of 120.2 and 134.7 Hz of the agostic and non-agostic C_β–H entities, respectively [90]. Hence, pronounced Ti···H_β interactions and C–H bond activation can be clearly ruled out in case of **17**. The same is true for **17a** showing a rather normal $^1J_{CH}$ coupling constant of 128 Hz for the agostic C–H moiety [91]. Hence, establishment of strong agostic interactions is prevented by the hinderance of negative hyperconjugative delocalization. However, in cases where direct M ← N π-donation is suppressed, delocalization of the M–N bonding electrons can occur in a "manner that closely resembles the delocalization of the Ti–C electron density in corresponding metal alkyl complexes" [92]. Hence, pronounced agostic interactions were found in aminoboranes such as $Cp_2Zr(Cl)NH_2BH_3$, $Cp_2Zr(H)NH_2BH_3$ [93], or $Cp_2TiNH_2BH_3$ [92]. We finally note that the "agostic protons" in d^0 amido complexes **17** and **17a** are characterized by small positive atomic charges and a downfield ^{1}H NMR chemical shifts ($\delta(^1\mathrm{H}) = 6.77$ and 2.45 ppm, respectively). These downfield shifts of the agostic protons were found to be even larger in related titanium agostic amido systems showing shorter Ti···H distances [90]. These observations are therefore in clear conflict with the assumed hydridic character of the agostic proton in the BG model [4, 8, 52]. Indeed, we will outline later (see Sect. 5.3) that the chemical shift of the agostic proton cannot be related to its atomic charge and is mainly controlled by the presence and extent of an opposing local Lewis-acidic site in the valence shell of the neighboring metal atom.

4.3.2 β-Agostic Silylamido and Silylmethyl Complexes

We outlined above that the $p_\pi(C_\alpha)$–$\pi^*(C_\beta H_\beta)$ interaction is usually weak in metal alkyls due to the rather electropositive character of the H_β atom. However, negative hyperconjugative $\pi \rightarrow \sigma^*$ delocalization and reduction of the carbanionic character at the C_α atom can be enforced by two strategies: (1) in case of transition metal alkyls the nodal structure of the d-orbital involved in the M–C_α bond provides the proper symmetry to act as acceptor orbital to stabilize the M···H_β bonding (Fig. 7 and 8). Hence, the lacking electronegative nature of the H_β atom can be compensated in transition metal alkyls by the establishment of covalent M···H bonding. (2) On contrast, main group alkyls cannot establish truly covalent M···H interactions due to the lack of d-orbitals. As a consequence the $p_\pi(C_\alpha)$–$\pi^*(C_\beta H_\beta)$ interaction remains weak, and the destabilizing four electron $p_\pi(C_\alpha)$–$\pi(C_\beta H_\beta)$ interaction dominates (Scheme 5).[8] Alternative strategies developed to stabilize carbanions therefore generally rely on the introduction of second-row substituents at the α-position which stabilize carbanions more efficiently than their first-row counterparts (see for example [72]). Especially, α-silyl substitutents are frequently used in this respect as they provide a polarizable and electropositive Si atom offering low-lying σ^* orbitals which can support negative hyperconjugation (see for example [72]).

This mechanism does not afford (but may be assisted) by secondary closed shell M···H interactions. As a consequence, no significant C–H activation has ever been reported for Li agostic alkyls. This is elegantly demonstrated by the benchmark system **8** which had been originally classified as a so-called lithium agostic system by Papasergio et al. [94] due to its short Li···H_γ contacts of 2.769(6) and 2.320(6) Å [54, 55] (Fig. 13). Besides this geometrical criterion, Kaufmann et al. considered lithium agostic interaction in terms of σ(CH) → Li donation of electron

[8] As another consequence of the destabilizing four electron $p(C_\alpha)$–$\pi(C_\beta H_\beta)$ interactions, the ethyl anion is assumed to be less stable than the methyl anion; see [71].

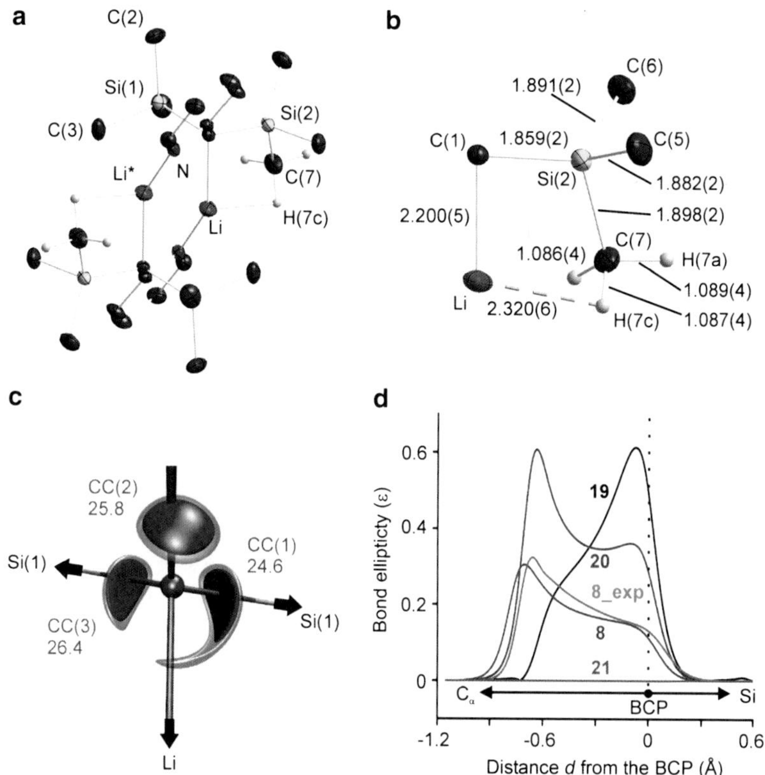

Fig. 13 (**a**) Molecular structure of $[2\text{-}(Me_3Si)_2CLiC_5H_4N]_2$ **8** based on a single crystal neutron diffraction study at 20 K; probability level 50%. (**b**) Agostic alkyl fragment and salient geometrical parameters; distances in Å and angles in (°); (**c**) the experimental valence charge concentrations at the carbanionic C_α atom denoted $CC_\alpha(1)$ and $CC_\alpha(2)$ (see Fig. 8 for the notation) have already merged into a single feature labeled CC(1) [$L(\mathbf{r})$ values at 17.5 and 19.5 eÅ^{-5}]. (**d**) Experimental and calculated bond ellipticity profiles (ε) along the C_α–Si_β bond path of **8** in comparison with CH_2SiH_2 **19**, $[CH_2SiH_3]^-$ **20**, and CH_3SiH_3 **21**. For a definition of ε, see Fig. 10

density [21]. However, closer inspection of the bonding characteristics of **8** reveals that all C–H bonds of the hydrogen atoms involved in the short Li···H contacts display standard C–H bond lengths (1.086(4)–1.089(4) Å) [54, 55] which do not signal any bond activation. Apparently, negative hyperconjugative delocalization in **8** is mainly driven by the electropositive character of the β-Si atom in the alkyl chain and only assisted by secondary closed shell Li···H contacts. As a consequence, we observe an acute $LiC_\alpha Si$ angle of 88.8(2)° a pronounced asymmetry of the C_α–Si (1.859(2) Å) and C_γ–Si (1.898(2) Å) bond lengths indicative of NHC delocalization. Calculations of the donor-free model system $LiCH_2SiH_2Me$ **18** show the same characteristic deformations of the alkyl backbone with a remarkably short Li···C_γ = 2.364 Å contact [$\angle LiC_\alpha Si$ = 88.0°, C_α–Si (1.834 Å), C_γ–Si

(1.946 Å)] [55]. Accordingly, (1) the coordination of the σ-donor ligand, (2) π-electron delocalization involving the aromatic pyridine ring, (3) secondary interactions such as Li···Li contacts or intermolecular Li···H contacts, and (4) other crystal-packing effects do not play a significant role for the alkyl group geometry in the Li agostic benchmark **8**. Thus, **8** was classified as a suitable experimental benchmark system to study the driving forces of Li···H–C bonding in the following.

In the charge density picture, the NHC effects are reflected by the merging of the valence charge concentrations $CC_\alpha(1)$ and $CC_\alpha(2)$ at the carbanionic C_α atom into a single feature denoted CC1 in Fig. 13c. Furthermore, NHC effects are evident from a complex bond profile showing a pronounced maximum in $\varepsilon(\mathbf{r})$ close to C_α and a shoulder close to the BCP [55]. The shoulder signals the development of $C_\alpha = Si$ double bond character in **8** which is, however, significantly reduced in comparison with the model system $CH_2 = SiH_2$ **19**. The pronounced maximum highlights once again the disturbance of the charge density distribution in the region of C_α by the carbanionic lone pair. However, in comparison with $[CH_2SiH_3]^-$ **20**, the carbanionic nature in **8** appears to be already significantly diminished.

This is possibly a direct consequence of the electron withdrawing character of the methyl substituent at Si_β which is involved in the formation of secondary Li···H_γ interactions. Hence, the lack of an appropriate d-orbital acceptor orbital at the metal has been compensated in **8** by the electropositive nature of the silicon atom in β-position of the alkyl backbone and via secondary Li···H_γ interactions. The chemical relevance of these secondary interactions can be, however, best demonstrated by the fact that the uncomplexed $[CH_2SiH_2Me]^-$ **18a** anion also reveals the typical geometrical signatures of negative hyperconjugative effects (e.g., asymmetric Si_α/β-C bond lengths of 1.773 and 1.953 Å, respectively) but a conformation of its alkyl backbone which is radically different from that in the corresponding lithium complex **18** [54]. Indeed, the position of the methyl group in the anion **18a** is energetically favored in an anti-orientation toward the carbanionic lone pair, and the C–Si–C angle (125.4°) is widened by more than 17° relative to **18**. Hence, the conformation of the alkyl backbone in our theoretical and experimental model complexes **8** and **18** critically depend on the presence of secondary Li···H_γ or Li···C_γ interactions [54, 55]. However, due to the electrostatic nature of these Li···H_γ bonds, no significant C_γ–H_γ activation can be observed and they mark the borderline case between covalent agostic interactions and purely electrostatic closed shell M···H contacts. This is also revealed by the flat and remarkably low charge density distribution in the Li···H_γ contact region (Fig. 5). These findings therefore relativize earlier statements in the literature concluding on the basis of theoretical studies that Li···H–C interactions might account for 40% of the valence shell electron density for the lithium atom [95].

Similar conclusions were also derived for the homoleptic bis(trimethylsilyl) amides and the bis(trimethylsilyl)methyl lanthanide complexes $Ln\{N(SiMe_3)_2\}_3$

and Ln{CH(SiMe$_3$)$_2$}$_3$ by Perrin et al. [96]. In these pyramidal complexes, γ-agostic M···H–C interactions were assumed to be essential to saturate the otherwise empty coordination sphere of the large lanthanide ions. However, like in ionic Li-agostic systems *all calculations show an elongation of the β-Si–C bond, while the γ-C–H bonds relatively close to La are not elongated.* Perrin et al. therefore also concluded that *the elongation of the Si–C bonds is interpreted as coming largely from a delocalization of the density in the sp^3 hybrid of the C centers near La into the β-Si–C bonds. This delocalization, which is also known as negative hyperconjugation, is especially important for silyl groups due to the relatively low-lying energy of the antibonding $\sigma^*_{Si–C}$ orbitals* [96]. A similar situation appears to hold in general for the rather ionic agostic complexes of lanthanides: neutron-diffraction studies of [Cp*Y(OC$_6$H$_3$Bu$^t_2$)CH(SiMe$_3$)$_2$] and [Cp*La{CH(SiMe$_3$)$_2$}$_2$] have revealed rather normal C–H bond lengths [17, 97]. In this case, Ln···Si–Me interactions were proposed to *clearly take precedence over α-C–H···Ln and γ-C–H···Ln interactions in stabilizing the lanthanide center.* However, as discussed for Li alkyls, NHC appears again to be the real driving force behind these deformations, and may extend to the case of *bimetallic*, so-called *polyagostic* lanthanide aluminate complexes such as [Ln(AlR$_4$)$_n$] (e.g., Ln = Sc, Sm, Nd for $n = 3$ and Ln = Yb for $n = 2$; R = Me, Et, ^{i}Bu) [98]. Hence, in contrast to d-block metals truly covalent M···H–C interactions seem to be less favored by ionic main group and lanthanide complexes. However, the increased polarity of a coordinating Si–H bond appears to be already sufficient to yield strongly activated β-agostic Ln···(Si–H) interactions – for example in ansa-bridged rare earth disilyamido complexes [99]. Hence, introduction of second-row substituents (especially silyl substituents) at the α-position of carbanionic ligands efficiently supports negative hyperconjugation which can be further enhanced by additional of M···H–X interactions (X = C, Si).

5 Control Parameters (II): Local Lewis-Acidic Centers at the ML$_n$ Fragment

5.1 Ligand-Induced Charge Concentrations in d^0 Transition Metal Alkyls

As demonstrated in the previous section, even secondary M···H contacts (e.g., in carbanionic lithium complexes) might assist negative hyperconjugative delocalization of the Li–C electron pair and thereby enhance alkyl group deformations. However, what are the electronic driving forces in transition metal alkyls that stand behind the formation of true covalent bonding between transition metal atoms and appended C$_β$H$_β$ moieties?

To answer this question, we should first find an explanation for the *presence/absence* of an agostic interaction in the 12 VE species $EtTiCl_3(dmpe)$ **2** and its four-coordinate (8 VE) parent complex $EtTiCl_3$ **3**, respectively (Fig. 1). The electronic structures of both TiEt fragments are closely related since both **2** and **3** display polar Ti–C bonds with noticeable carbanionic character [11]. However, canting of the ethyl group to allow a positive and significant overlap between the doughnut-shaped density contours of the Ti ($4s/3d_{z}{}^{2}$) hybrid orbital and the agostic hydrogen atom is energetically favored only in case of the phosphine adduct **2**. Hence (1) steric congestion, (2) insufficient Lewis acidity, (3) a too high *VE count* (VE = 8), or (4) the lack of a suitable d-acceptor orbital can all be ruled out as reasons for the hindrance of the agostic interaction in $EtTiCl_3$. It is therefore important to search for the microscopic origin which counteracts the β-agostic interaction in **3**.

Inspection of the $L(\mathbf{r})$ contour maps of **3** reveals four pronounced *ligand-induced charge concentrations* (LICCs) opposing the four σ(Ti–X) bonds (X = C, Cl) at the ligand-opposed side of the metal. These LICCs are therefore denoted *trans*-CC (C) and *trans*-CC(Cl) (3×) in Fig. 9d, respectively [11, 100]. In addition, a fifth CC located on the Ti–C bond path can be identified as BCC(C) in analogy to the expression *bonded charge concentration* (*BCC*) used in main group chemistry [101]. Such a BCC(C) is a characteristic feature of covalent M–C bonding in early transition metal compounds – and has never been observed in case of the ionic metal alkyls M–R (where M is a Group 1 element, Be, Mg, or Al) (e.g. Fig. 8). As illustrated in Fig. 9a–c, the nature and occurrence of all these LICCs can be derived by a direct and rigorous interpretation of the wavefunction: *they are an integral part of the bonds formed using metal orbitals possessing d-character. It is then the different nodal structures of p and d wavefunctions that give rise to different atomic polarization patterns at main group atoms and transition metals* [100].

According to a natural bonding orbital (NBO) analysis [102], the highly polar Ti–C bond (Fig. 9b, c) shows 30% metal and 70% carbon character furnishing formally a $sp^{0.4}d^{3.9}$ hybridization at the Ti atom (19% s, 7% p, and 74% d character). The significant Ti(p) admixtures cause a *reduction* in the charge concentration [BCC(C)] in the $sp^{0.4}d^{3.9}$ lobe facing the C_α atom and an *increase* in the charge concentration [*trans*-CC(C)] in the opposing $sp^{0.4}d^{3.9}$ lobe at the far side of the metal (Fig. 9a, c). Formation of the polar σ(Ti–C) bond *in the NBO picture* is therefore paralleled by the establishment of three types of charge concentrations at the metal in the *charge density picture*: one pronounced *trans*-CC(C) in the ligand-opposed lobe of the $Ti(sp^{0.4}d^{3.9})$ hybrid orbital and a small ligand-directed charge concentration BCC(C) in opposite direction (Fig. 9a). Additionally, we observe a rather diffuse ring-shaped *cis*-CC(C) in the region of the doughnut-shaped density contours of the $Ti(sp^{0.4}d^{3.9})$ hybrid orbital (Fig. 9a). In the total charge density distribution (Fig. 9d), this rather diffuse *cis*-CC(C) merges with the three ligand-opposed charge concentrations *trans*-CC(Cl) of the chlorine ligands which dominate the $L(\mathbf{r})$ fine structure in the valence shell of Ti atom [100, 103].[9,10] *Hence, the presence of LICCs and hybrid orbitals involving* s, p, and d AOs *at the metal simply represents two sides of the same coin.*

Figure 9d illustrates that the four ligand-opposed charge concentrations appear to prevent *as stereochemically active centers* [100, 104, 105] the coordination of the $C_\beta H_\beta$ moiety in **3** (Fig. 9d). Hence, inspection of the $L(\mathbf{r})$ fine structure in **3** provides a visualization of the *trans*-influence displayed by the chloro ligands in the MO picture [11], which hinders the coordination of an agostic $C_\beta H_\beta$ moiety. Accordingly, the hindrance of β-agostic interaction in **3** might be interpreted in terms of a topological frustration: no presence of sterically accessible local Lewis-acidic centers at the metal center to accommodate the agostic $C_\beta H_\beta$ moiety. However, this "topological frustration" scenario can be overcome simply by removal of the Cl^- ligand opposite to the $C_\beta H_\beta$ unit. In that case, the $C_\beta H_\beta$ moiety is no longer electronically passivated by the *trans*-influence of any opposing σ/π-donor chloro ligand and can undergo an agostic interaction with the titanium center. Indeed, the charge density picture shows that removal of the chloro ligand *trans* to the $C_\beta H_\beta$ entity yields a local charge depletion (CD) zone at the near side of the metal next to the agostic C–H moiety. Hence, the formation of such a Lewis-acidic site initiates and enhances the agostic interaction in the $[EtTiCl_2]^+$ cation **6** (Fig. 9f). We note that such a charge depletion zone is not formed in the parent complex **3** even by enforcing an agostic geometry via a restrained geometry optimization (Fig. 9e). This clearly shows that the charge depletion zone in the $[EtTiCl_2]^+$ cation is not a consequence of the agostic interaction but rather its origin.

Inspection of the $L(\mathbf{r})$ contours in Fig. 14 also reveals the electronic and structural relationship between the agostic model systems $[EtTiCl_2]^+$ **6**, $[EtTiCl_2L]^+$ **22,** and $[EtTiCl_2L_2]^+[Cl^-]$ **2** ($L = PMe_3$, $L_2 =$ dmpe). According to this notation, the $EtTiCl_3$(dmpe) complex can be formally considered as an $[EtTiCl_2L_2]^+[Cl^-]$ ion pair. This is supported by the fact that the Ti–Cl(1) bond is one of the longest Ti–Cl bonds (2.4223(2) Å) ever reported which is significantly weakened in comparison to the ones displayed by $EtTiCl_3$ (2.195(3) Å; Fig. 1) [11]. We first note that the phosphine ligands in **22** and **2** coordinate approximately *trans* to the agostic hydrogen atom. Hence, they share the same d-orbital at the metal center with the

[9] Bader et al. have demonstrated that the negative Laplacian of the charge density distribution, $L(\mathbf{r}) = -\nabla^2\rho(\mathbf{r})$, determines where the charge density distribution is *locally concentrated* ($L(\mathbf{r}) > 0$) or locally depleted ($L(\mathbf{r}) < 0$). Accordingly, the $L(\mathbf{r})$ function can be used to resolve the shell structure for elements with $Z \leq 18$. However, the shell structure of the transargonic elements is not fully represented by the Laplacian. In general, the fourth, fifth, or sixth shell for elements of periods 4–6, respectively, is not revealed in the Laplacian. As a convention, Bader et al. suggested that the outermost shell of charge concentration (CC) of an atom (*second* shell of CC of the carbon atoms and *third* shell of CC of the nickel atom) represents its (effective) valence shell charge concentration (VSCC).

[10] We note that the higher polarity of the Ti–Cl bond is signaled in the charge density picture by the vanishing of the corresponding BCCs in the total charge density distribution in the Ti–Cl bonding region and by a smaller magnitude of *trans*-CC(Cl) vs. *trans*-CC(C).

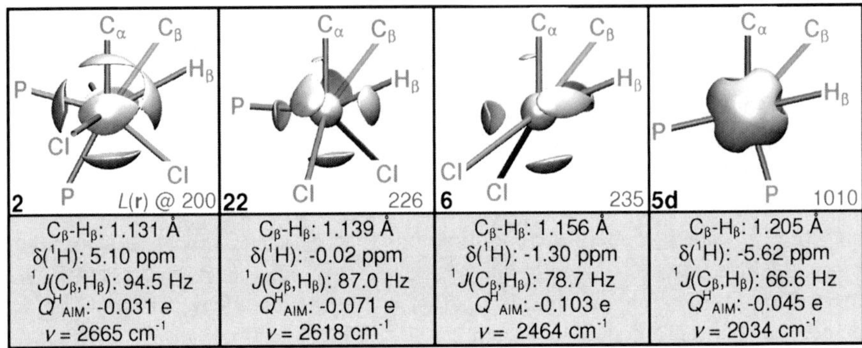

2
L(r) @ 200
Cβ-Hβ: 1.131 Å
δ(¹H): 5.10 ppm
¹J(Cβ,Hβ): 94.5 Hz
Q^H_AIM: -0.031 e
ν = 2665 cm⁻¹
22
226
Cβ-Hβ: 1.139 Å
δ(¹H): -0.02 ppm
¹J(Cβ,Hβ): 87.0 Hz
Q^H_AIM: -0.071 e
ν = 2618 cm⁻¹
6
235
Cβ-Hβ: 1.156 Å
δ(¹H): -1.30 ppm
¹J(Cβ,Hβ): 78.7 Hz
Q^H_AIM: -0.103 e
ν = 2464 cm⁻¹
5d
1010
Cβ-Hβ: 1.205 Å
δ(¹H): -5.62 ppm
¹J(Cβ,Hβ): 66.6 Hz
Q^H_AIM: -0.045 e
ν = 2034 cm⁻¹

Fig. 14 Calculated ligand-induced changes of the $L(\mathbf{r})$ fine structure at the titanium atom along the coordination series $[EtTiCl_2]^+$ **6**, $[EtTiCl_2L]^+$ ($L = PMe_3$) **22**, $[EtTiCl_2L_2]^+[Cl^-]$ (L_2 = dmpe) **2**, in comparison with the polarization pattern of the agostic Ni d^8 cation $[(DCpH)Ni(d^tbpe)]^+$ **5d**. Note the increasing ^1H down-field shifts with increasing shielding of the metal atom by ligand-induced charge concentrations opposite to the $C_\beta H_\beta$ moiety. $L(\mathbf{r})$ contour values are specified in (eÅ^{-5}). The coordination vectors to the ligand atoms are only indicated for orientation purposes and are not true to scale; C–H streching frequencies of the agostic moieties are specified

agostic hydrogen atom via a multicenter interaction. This multicenter interaction involving the same metal d-orbitals is reflected by calculated J(P,H) coupling constants of ca. 5.1 Hz in **2** [$J(P,H)_{exp}$ = 6.6 Hz] and **22**. We note already at this stage that these coupling constants approach values around 30 Hz in the corresponding agostic late transition metal complexes **5a–d** in line with their stronger $P\cdots M\cdots H_\beta$ 3c2e interaction. This multicenter interaction is also reflected in the charge density picture of **2** and **22** by the presence of two charge concentrations opposite to the Ti–H and Ti–P bonds. Again, we can classify these features as ligand-opposed charge concentrations *trans*-CC(H) and *trans*-CC(P) (Fig. 14). Hence, the $L(\mathbf{r})$ polarization patterns of **22** and **2** are highly related. Nevertheless, the titanium atom in the dmpe complex **2** appears to be slightly larger shielded. This is a consequence of the additional P–Ti coordination approx. *trans* to the Ti–C_β direction which results in a partial filling of the charge depletion zone facing the C_β atom which apparently leads to a reduced agostic bond strength.[11] This is reflected in a rather small D_{112} value of ca. 8 kJ mol^{-1} (35 kJ mol^{-1} in **6**) [15] and an only modest C–H bond activation in comparison with **6** and **22** (Fig. 14). The increasing shielding of the metal center in the series **6** → **22** → **2** by an increasing number and extent of the LICCs correlates with a stepwise reduction of the local Lewis acidity in the $Ti\cdots C_\beta H_\beta$ bonding region. This is not only reflected by a reduction in the C_β–H_β bond activation in the series **6** → **22** → **2** but also mirrored in drastic changes of the $\delta(^1H)$ chemical shifts of

[11] As a consequence of the phosphine coordination *trans* to the C_β atom, a weakly defined (3, −1) saddle point is formed in the $L(\mathbf{r})$ pattern at the titanium atom along the Ti–C_β vector. In the experimental charge density distributions, however, a subtle (3, +1) CD zone is preserved opposite to the C_β atom.

the agostic protons (Fig. 14). This surprising phenomenon will, however, be discussed in greater detail in Sect. 5.3.

It is therefore the higher coordination number and sterical congestion of **2** in comparison with **3** which prevent the coordination of a σ/π chlorine donor *trans* to the agostic CH moiety which again would hinder the establishment of agostic interactions via its *trans*-influence. As another consequence of the phosphine coordination in **2**, the Ti–C_α bond becomes significantly weakened relative to the one in **3** [Ti–C = 2.1537(8) Å in **2** and 2.090(15) Å in **3**] due to the geometrically enforced *trans*-coordination of the chelating ligand. Ironically, the canting of the ethyl group which initiates the agostic interaction in **2** is thus facilitated by the high steric congestion in **2**.

Hence, in contrast to the original suggestion by Brookhart and Green [4, 8], it is therefore not the *global* Lewis acidity of the metal center, but the presence of *locally induced sites* of increased Lewis acidity which determines the extent of agostic interactions [10, 11]. As a matter of fact, no agostic complex has been reported so far where agostic $C_\beta H_\beta$ moieties are facing a domain of local charge concentration at the metal coordination site.[12] The strongest $M\cdots H_\beta C_\beta$ agostic interactions are indeed displayed when either the C_β (typically in d^0 complexes) or H_β atoms (typically in d^n late transition metal alkyls) are directly facing local Lewis-acidic centers at the metal center. We finally note that this concept was systematically investigated by a controlled manipulation of the metal polarization in case of a series of complexes $[EtTiCl_2L]^+$ [where L = (1) a strong *π-acceptor* (CO or PF_3), (2) a weak *π-acceptor* (PMe_3), (3) a *σ-donor* (H^-, CH_3^-, or NMe_3), or (4) a *π-donor* ligand (Cl^-, F^-, or OMe_2)]. In all cases, the extent (or hindrance) of the agostic interactions could be related to the local polarization pattern at the metal atom [11].

5.2 The Fine Structure of the Laplacian in d^n Transition Metal Alkyls

As outlined above, the strength of agostic interactions in d^0 early transition metal alkyl complexes is mainly controlled by (1) the presence of a local Lewis-acidic center at the metal atom in the coordination area of the agostic $H_\beta C_\beta$ moiety, (2) the extent of negative hyperconjugative delocalization of the M–C bonding electrons, and (3) to a smaller degree by σ(M ← H–C) donation [10, 11, 90]. *Accordingly, the bonding between the metal atom and the ethyl group in d^0 transition metal alkyls is effectively established by one electron pair/molecular orbital* [10, 15]. However, our MO analysis (Sect. 3) reveals that the electronic structures of agostic moieties change significantly when we move on to electron-rich agostic transition metal complexes.

---

[12] Hence, the presence of local Lewis-acidic sites in the coordination region of agostic $C_\beta H_\beta$ moieties might reflect the presence of d-acceptor orbitals which accommodate the $C_\beta H_\beta \rightarrow M$ donation in the MO picture.

^{t}Bu ^{t}Bu Ni HBF$_4$ ^{t}Bu ^{t}Bu Ni ⊕ H H ⊖BF$_4$

olefine: *ethylene (**4a**)*
*nbe (norbornene, **4b**)*
*D[H]$_2$Cp (dicyclopentene, **4c**)*
*DCp (dicyclopentadiene, **4d**)*

alkyl: *ethylene (**5a**)*
*nbeH (norbornyl, **5b**)*
*D[H]$_2$CpH (dicyclopentyl, **5c**)*
*DCpH (dicyclopentenyl, **5d**)*

Scheme 7

Here M → L π back donation, which furnishes simultaneously *covalent* M–C$_\alpha$ and M–H$_\beta$ bonding, becomes dominant. This should be also reflected in the charge density picture. Especially, the local fine structure of $L(\mathbf{r})$ should differ significantly in d^0 and d^n ($n \geq 2$) metal complexes due to the presence of the free d-electrons in the latter case. This can be illustrated in case of agostic Spencer-type d^8 nickel-alkyl cations (**5a–5d**) which can be elegantly obtained simply by protonation of the corresponding d^{10} olefin nickel complex (η^2-C_2H_4)Ni(d^{t}bpe) **4a** (Scheme 7) [25, 26].

Unfortunately, severe disorder of [EtNi(d^{t}bpe)]$^+$[BF$_4$]$^-$ precludes Spencer's parent complex from experimental charge density studies. Nevertheless, well-ordered agostic systems (**5b–5d**) could be obtained by a systematic increase in the steric rigidity of the olefin backbone [26]. We note that in the range of the experimental errors, all models **5a–5d** display isostructural agostic NiC$_\alpha$C$_\beta$H$_\beta$ fragments which are characterized by highly acute MCC valence angles of ca. 75°, short C–C bond lengths of less than 1.50 Å, and highly activated C$_\beta$–H$_\beta$ bonds of ca. 1.20 Å length (Fig. 2 and Table 3).

The beauty and simplicity of Spencer's approach to render an olefin complex into an agostic alkyl species allowed us to inspect the changes of the electronic structure at the start and end point of this transformation in detail (Figs. 2 and 6; Table 5). Inspection of the calculated $L(\mathbf{r})$ contours of the parent nickel olefin complex **4d** and its agostic congener **5d** in the NiC$_2$ plane (Fig. 6) reveals distinct charge density rearrangements. Most important, the C=C double bond character becomes significantly reduced upon protonation, accompanied by a reduction in the charge density at the C–C BCP in **5d** [$\rho(\mathbf{r})_{exp} = 1.77(2)/\rho(\mathbf{r})_{calc} = 1.70$ eÅ^{-3}] in comparison with its corresponding olefin complex **4d** [$\rho(\mathbf{r})_{calc} = 1.89$ eÅ^{-3}] or our experimental benchmark (C_2H_4)Ni(d^{t}bpe) **4a** [$\rho(\mathbf{r})_{exp} = 2.107(9)$; $\rho(\mathbf{r})_{calc} = 1.92$ eÅ^{-3}]. However, the cations **5a** and **5d** still retain a distinct double bond character as witnessed by ellipticity profiles showing pronounced ellipticity values larger than zero in the full C–C valence region – in clear contrast to the agostic d^0 species EtTiCl$_2^+$ (Fig. 10) and EtTiCl$_3$(dmpe) [10]. This conforms to the fact that the protonation causes only a modest elongation of the former olefinic CC double bond in **4a** and **4d**. The differences in the electronic structures of the olefin species and their corresponding agostic congeners are, however, better documented by the differences in the $CC_\alpha(2)$ and $CC_\beta(2)$ values in the valence shell of the C_α and C_β atoms, respectively (Fig. 6) [106].[13] The short Ni–H_β bond length of 1.671(9) [1.653] Å classifies **5d** as an agostic benchmark system which is at a late stage of the β-elimination pathway and therefore close to the *cis*-ethene hydride form (theoretical values obtained by DFT calculations are given in square brackets in the following). The already remarkable strength of the Ni–H bond is reflected by a significant electron density accumulation at the Ni–H_β BCP of 0.553(4) [0.569] $eÅ^{-3}$ which approaches the values at the Ni–C_α BCP (0.680(9) [0.735] $eÅ^{-3}$) – our internal standard of a covalent nickel-ligand bond. However, (1) the large bond ellipticities, ε, (2) the small negative value of the total energy density, $H(\mathbf{r})$, and (3) the high density accumulation at the RCP (0.533 [0.507] $eÅ^{-3}$) inside the $[NiC_\alpha C_\beta H_\beta]$ fragment still reveals an unstable Ni···H bond path topology (Table 5). As outlined in Sect. 3 this observation is a natural consequence of pronounced L → Ni σ donation which gives rise to a large charge density accumulation inside the agostic moiety of **5d** and thus represents a signature of the delocalized nature of the agostic interaction.

As a consequence of the predicted Ni → L π back donation (Sect. 3), both experimental and theoretical charge density studies reveal a significantly elongated C_β–H_β bond distance of 1.20(1) Å [1.205 Å]. Furthermore, the charge density accumulation at the C_β–H_β BCP is clearly reduced $\rho(\mathbf{r})_{C\text{–}H} = 1.33(3)$ [1.387] $eÅ^{-3}$ (Table 5) relative to the weakly activated C_β–H_β bond in the d^0 alkyl (e.g., $EtTiCl_3(dmpe)$ **2**; C_β–H_β = 1.13 [1.131] Å; $\rho(\mathbf{r})_{C\text{–}H} = 1.54(5)$ [1.684] $eÅ^{-3}$) [11].

Interestingly, the $L(\mathbf{r})$ pattern at the metal in the olefin complexes (**4a**, **d**) and their agostic counterparts (**5a**, **d**) show only tiny differences (Fig. 6). This hints for the presence of closely related electronic structures in both types of complexes. To analyze the polarization pattern at the Ni atom, we first define the standard local coordinate system of a square-planar complex with the *x*- and *y*-axes roughly pointing to the ligand atoms (C_α, H_β) in the agostic species (Fig. 6). Due to compatibility reasons, we will use the same coordinate settings also for the approximately trigonal–planar olefin complexes **4a** and **4d**. The occurrence of four depletion zones of approximately equal $L(\mathbf{r})$ magnitudes along the *x*- and *y*-coordinate axes in **4a**, **4d**, and **5d** suggests the depopulation of the $d_{x^2-y^2}$ orbital relative to the other four d-orbitals in the formally d^8/d^{10} configurated nickel alkyl and nickel olefin complexes, respectively (Fig. 6). This could be verified by the experimentally determined $P(d_{x^2-y^2})$ population parameters based on the multipolar models of **4a** [$P(d_{x^2-y^2}) = 0.89(3)$ e] and **5d** [$P(d_{x^2-y^2}) = 1.62(2)$ e] [26, 40]. This decrease in d-orbital population reflects the $Ni(d_{x^2-y^2})$ → L π back donation in the π* orbitals

[13] We note that the relative magnitudes of $CC_\alpha(1)$, $CC_\alpha(2)$, and $CC_\beta(2)$ are similar in the experimental and calculated models. However, the experimentally determined charge concentrations appear to be larger than their respective theoretical ones. This might be due to the fact that core contraction/expansion phenomena have not been taken into account during the multipolar refinements [106].

of the olefin ligand in **4a, d** and the agostic alkyl ligand in **5d** (Fig. 7). Since the four CD zones in the charge density picture are directly connected with the depletion of the metal $d_{x^2-y^2}$ orbital in the MO picture, the angle between these CD zones is constrained and dictates the position of all ligand atoms in a key and lock scenario. Indeed, the metal-directed VSCC of each ligand (L) atom faces one of the four CD zones (representing local Lewis-acidic centers) at the metal atom (Fig. 6).

We conclude that the $L(\mathbf{r})$ fine structure and the MO model represent two sides of the same coin – like in case of the d^0 complexes. In d^0 complexes, the CD zone [a (3, +1) CP in $L(\mathbf{r})$] is shifted closer to the $M\cdots C_\beta$ directrix (Figs. 3, 4, 9, and 14) while in d^n configurated transition metal complexes ($n \geq 2$) the local Lewis-acidic site is typically facing the H_β atom (see for example Fig. 6). Apparently, the variance in the $L(\mathbf{r})$ fine structure is due to the dominance of *negative hyperconjugation* in early d^0 agostic species, whereas Ni $\rightarrow$ L π *back donation* dominates the electronic structures of the d^n congeners (Fig. 7). In the former case, the d-orbital involved in M–C bonding [HOMO in Fig. 7 (top)] also accommodates the $M\cdots H_\beta$ interaction. Accordingly, this prevents the formation of a pronounced charge depletion zone along the $M\cdots H$ bond directrix. In the latter case, however, the corresponding $M–C_\alpha/M–H_\beta$ bonding orbital [HOMO-4 in Fig. 7 (middle)] displays a nodal plane in the alkyl moiety and furnishes as a consequence of the Ni $\rightarrow$ L π *back donation* two depletion zones along the $M–C_\alpha$ and $M–H_\beta$ bond. Hence, analysis of the fine structure of the Laplacian might be used to quantify the extent of the individual driving forces of agostic interactions in late and early transition metal complexes.

5.3 *Valence Shell Fine Structure, C–H Bond Activation, and* ^{1}*H NMR Properties*

In this section, we will show how the differences in the valence shell fine structure of agostic d^0 and d^n ($n \geq 2$) transition metal complexes correlate with the extent of C–H activation and the ^{1}H NMR chemical shifts of the agostic protons. If not specified otherwise, all NMR properties were obtained from single-point calculations using the PBE0 hybrid functional [107] in combination with the TZ2P basis set and the spin–orbit ZORA formalism [35] [$\delta = \sigma(\mathrm{TMS})\text{-}\sigma$; with $\sigma^{calc}(\mathrm{TMS}) = 31.59$ ppm]. Geometry optimizations and relaxed PES scans were performed on the BP86/scalar-ZORA/TZ2P level of approximation.

We have outlined in the previous sections that β-H activation is linked with a significant redistribution of the bonding electron density upon the development of a $M\cdots H_\beta–C$ interaction. This electronic redistribution that leads to an activation of the agostic C–H bonds is usually reflected by reduced $^1J_{CH}$ coupling constants and an upfield shift of the agostic proton relative to an uncoordinated C–H group on the ^{1}H NMR scale [4, 8, 10, 76, 108]. Especially the former criterion of reduced $^1J_{CH}$ coupling constants has been successfully advanced in numerous theoretical studies to quantify the extent of C–H bond activation by relating the $^1J_{CH}$ coupling constants with C–H bond distances (see for example [90, 108–110]). In experimental studies, the situation might be more complex, especially when fast dynamical processes (on the NMR time scale), e.g., the rotation of the terminal methyl groups in β-agostic compounds only provide data of averaged $^1J_{CH}$ coupling constants. Indeed, the $^1J_{CH}$ value observed for **2** is 126.8 Hz [15], which is far above the value typically assumed for agostic C–H bonds (75–100 Hz [4]). This is due to the fluxionality of the system yielding, even at −90°C, an averaged coupling constant in good agreement with the calculated averaged $^1J_{CH}$ values for the agostic C–$H_β$ moiety of $^1J_{CH}$ = 129.3 Hz; individual values are 94.5 Hz for the agostic C–$H_β$ bond and 146.3 and 147.2 Hz for its non-agostic congeners in the terminal methyl group. Figure 14 shows that the increasing strength of the agostic interaction in the compound series **2** → **22** → **6** → **5d** nicely correlates with the corresponding reduction of the computed $^1J_{CH}$ values (Hz): **2** (94.5) → **22** (87.0) → **6** (78.7) → **5d** (66.6).

In case of ^{1}H chemical shifts, the situation becomes further complicated by the fact that the chemical shifts of β-agostic protons do not conform with the original assumption that agostic hydrogen atoms are always characterized by an upfield shift in the proton NMR [111]. The computed ^{1}H chemical shifts in the series **2** (+5.10) → **22** (−0.02) → **6** (−1.3) → **5d** (−5.62) (Table 1; Fig. 14) are also in conflict with the earlier assumption that *for d^0 systems, resonances due to the agostic hydrogens normally do not occur at higher fields than 0 ppm* [8]. This is nicely underpinned by the experimental NMR studies on the β-agostic d^0 complexes $[Cp^*Zr(^iBu)\{N(Et)CMeN(^tBu)\}]^+$ ($\delta(^1H)$ = −0.27 ppm; $^1J_{CH}$ 96 Hz) [112] and $[Cp_2Ti(CH_2CHMeCH_2CHMe_2]^+$ ($\delta(^1H)$ = −3.43 ppm; $^1J_{CH}$ 87.2 Hz) [113] which might represent suitable experimental model systems of **22** and **6**. Hence, the large variance of the ^{1}H chemical shifts in agostic compounds is more complex and not simply related to the d-electron count of the metal and warrants a detailed analysis by charge density methods.

In the first step of this analysis, the individual contributions to the ^{1}H NMR shielding tensor in various agostic reference systems were studied. Since spin–orbit coupling effects were treated explicitly in the calculations, the isotropic shielding values consist of three contributions: the diamagnetic (σ^d), paramagnetic (σ^p), and spin–orbit (σ^{so}) term. As outlined above, agostic d^0 transition metal alkyls such as **2** or the amido complex **17** are not necessarily characterized by an upfield shift of the ^{1}H NMR signal of the agostic protons [15, 90]. As outlined above, this is in conflict with the general presumption in literature that agostic protons are significantly shielded due to their proximity to the metal center and their partial hydridic character. Indeed, in case of such a hydridic character, the assumed charge density accumulation at the agostic β-hydrogen atom should result in an increased diamagnetic shielding contribution, σ^d, corroborated by an upfield shift of the ^{1}H signal. This generally accepted assumption is, however, not supported by our experimental and theoretical charge density analyses and NMR studies. Accordingly, our experimental d^0 model complex $EtTiCl_3(dmpe)$ **2** and its agostic d^8 congener $[(DCpH)Ni(d^tbpe)]^+[BF_4]^-$ **5d** are both characterized by almost vanishing atomic charges at

Table 6 ^{1}H NMR chemical shifts (δ) (ppm) together with the diamagnetic (σ^d), paramagnetic (σ^p), and spin–orbit (σ^{so}) shielding contributions of the agostic protons in the agostic benchmarks **2** and **5a**

$\angle MC_\alpha C_\beta$ (°)	d M–H (Å)	d C_β–H (Å)	δ (ppm)	σ^d (ppm)	σ^p (ppm)	σ^{so} (ppm)	Q^H_{AIM} (e)
5a							
76.2[a]	1.682	1.196	−3.78	28.16	5.74	1.47	−0.053
75.b[b]	1.634	1.215	−6.06	28.21	7.28	2.16	−0.052
73.9[c]	1.589	1.241	−9.59	28.02	9.81	3.36	−0.047
2							
87.2[d]	2.200	1.125	4.77	27.47	−0.59	−0.06	−0.019
85.1[e]	2.110	1.131	5.10	27.64	−1.08	−0.08	−0.031
82.5[f]	2.000	1.139	5.48	28.00	−1.80	−0.10	−0.043

[a]Relaxed PES scan with variable Ni–P_{trans} distances of 2.0 Å
[b]Relaxed PES scan with variable Ni–P_{trans} distances of 2.156 Å (equilibrium geometry)
[c]Relaxed PES scan with variable Ni–P_{trans} distances of 2.4 Å
[d]Relaxed PES scan with variable Ti–H distance of 2.2 Å
[e]Relaxed PES scan with variable Ti–H distance of 2.11 Å (equilibrium geometry)
[f]Relaxed PES scan with variable Ti–H distance of 2.0 Å

the agostic hydrogen atoms ($Q_{AIM}{}^H = 0.13$ [−0.031] in **2**) and $Q_{AIM}{}^H = -0.01$ [−0.045] in **5d** (Table 1). The pronounced downfield/upfield ^{1}H chemical shifts of **2** and **5d**, respectively, point therefore to a quite different physical origin. The expression "hydridic shift" which is often used in literature to correlate the chemical shifts with the charge density accumulation at the hydrogen atoms bonded to the metal appears therefore to be misleading, and its origin needs a more sophisticated treatment (see for example [114]). This result is therefore in clear conflict with the BG model, assuming that agostic hydrogen atoms carry a negative charge which consequently should lead to an upfield ("hydridic") ^{1}H NMR shifts of the agostic protons [4, 8].

Further analysis of the individual diamagnetic and paramagnetic contributions to the isotropic shielding shows, however, that the ^{1}H NMR shifts for the agostic protons depend mainly on the sign and magnitude of the paramagnetic contribution, σ^p. Note that in case of late transition metal complexes like **5d** also the spin–orbit (σ^{so}) term might contribute to the chemical shift of the agostic proton. Nevertheless, the σ^{so} term displays the same sign as the σ^p term in all our benchmark systems (Tables 1–3) and will therefore not be considered explicitly in the following. Tables 1–3 also show that the magnitude and sign of σ^p do not depend on the presence or absence of free d-electrons as exemplified by the large σ^p values for our cationic d^0 model system $[EtTiCl_2]^+$ (**6**) ($\sigma_{calc}{}^p = +4.3$ ppm) and its corresponding d^8 system $[DCpHNi(d'bpe)]^+$ **5d** ($\sigma_{calc}{}^p = +6.09$ ppm), while the neutral d^0 benchmark **2** displays a negative $\sigma_{calc}{}^p$ value of −1.08 ppm in line with a downfield ^{1}H shift and the observation of a *negative* isotopic perturbation of resonance (IPR) in experimental NMR studies [15].

Table 6 reveals that the charge at the agostic hydrogen atom also remains rather invariant upon reduction of the M···H distances, and the same is consequently true for the diamagnetic shielding contribution. Indeed, the upfield shift in the d^8

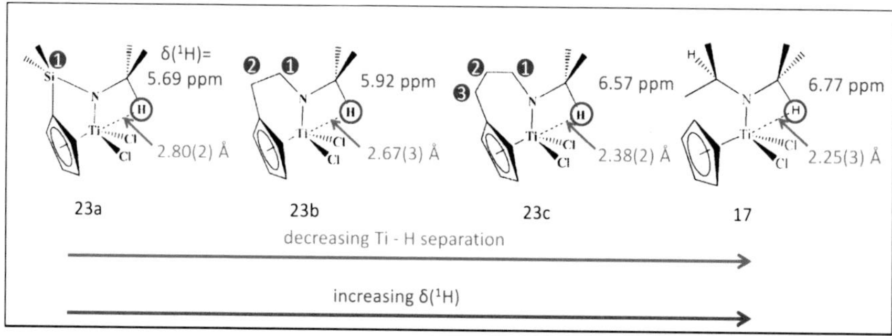

Fig. 15 Correlation between the calculated ^{1}H NMR chemical shift of the agostic proton at different Ti···H distances in [{η^5-$C_5H_4SiMe_2N$(*i*Pr)}$TiCl_2$] (**23a**; [28]), [{η^5-$C_5H_4(CH_2)_2N$(*i*Pr)}$TiCl_2$] (**23b**; [29]), [{η^5-$C_5H_4(CH_2)_3$ N(*i*Pr)}$TiCl_2$] (**23c**; [29] and **17** [90]

complex **5a** is basically due to the pronounced σ^p term which increases with decreasing M···H distances (Table 6). The opposite trend (downfield shift upon M···H reduction) is reflected in the calculated NMR properties of the d^0 alkyl species **2** and in experimental NMR studies of the d^0 amido complexes **17** and **23a–c** (Fig. 15) [90]. One might conclude at this stage that it is the metal to hydrogen distance which mainly controls the magnitude of the σ^p contribution. But what are the electronic parameters which control its sign?

The latter question (sign of the σ^p contribution term) can be answered by inspection of the VSCC pattern of our d^0 titanium model systems **2**, **6**, and the d^8 nickel complex **5d** in Fig. 14. This reveals a clear topological trend: large upfield shifts of the agostic proton are only observed in a topological scenario where the β-agostic hydrogen atom is facing a *charge depletion zone* [or more precisely a (3, +1) CP in the valence shell of charge concentration] at the metal center (e.g., in **5d**; Fig. 14d). In contrast, pronounced downfield shifts are found where the β-agostic hydrogen atom is close to a *charge concentration* or (3, −3) CP (e.g., in **2**; Fig. 14a) at the metal. Also the theoretical model systems **6** and **22** follow this trend and mark intermediate cases between strong and weak agostic interactions (Table 1). Indeed, in the equilibrium geometry of **6** the agostic H_β atom is facing a *saddle point* [or (3, −1) CP in the $L(\mathbf{r})$ maps], while the C_β atom moiety is facing a charge depletion zone. The increasing shielding of the titanium atom by LICCs in the series **6** → **22** → **2** (Fig. 14) is therefore matched by an increasing downfield shift of the ^{1}H signals in the proton NMR: **6** [−1.3 ppm] → **22** [−0.02 ppm] → **2** [+5.1 ppm]. Apparently, the paramagnetic contributions from the metal fragments to the ^{1}H chemical shift might be "shielded" by local charge concentrations facing the agostic $C_\beta H_\beta$ moiety. We finally note that also in organic nitranions, a relationship between the topology of the Laplacian and changes in the ^{15}N chemical shifts was observed before [115]. Hence, correlations between ground state electron densities and chemical shifts might not be uncommon – an aspect which clearly warrants further explorations.

6 Conclusion

Current bonding concepts of agostic interactions were summarized to provide a linkage between the molecular orbital view and its complementing electron density picture. Originally, agostic interactions were considered in the BG model to arise from the donation of electron density from a C–H group of an appended alkyl group into a vacant orbital of the electron-deficient metal atom, resulting in a 3c–2e M···H–C moiety [4, 8]. The poor predictive power of the original BG model, however, stimulated intensive research to identify the microscopic control parameter of agostic interactions and the activation of C–H bonds.

The current bonding concepts, which were summarized and discussed in this chapter, consider the agostic interaction in early transition metal alkyl complexes as a *covalent interaction which can be understood only in terms of a M–C_α bonding orbital that is delocalized over the entire agostic alkyl moiety. The reduction of the MCC valence angle in case of β-agostic alkyls thus allows the metal atom to establish bonding interactions with the β-C and its appended H atom* [15]. In the charge density picture, the *delocalized* nature of the β-agostic interaction is often reflected by a missing M···H_β bond path or a bond catastrophe scenario. Hence, the existence of an agostic bond can usually not be proven by the triplet of concomitant topological objects: a BCP, a BP, and an IAS as suggested by Popelier [38].

The reduced antibonding character of the C–C antibonding $\pi_{z'}$ orbital in agostic ethyl ligands relative to the $[C_2H_5]^-$ anion among other observations led SGM to conclude that agostic stabilization in transition metal alkyls *arises from negative hyperconjugative delocalization of the M–C bonding electrons* [10, 11, 55]. In this respect, closed shell or covalent M···H_β interaction support the extent of negative hyperconjugation in a related way as the replacement of H_β by a more electronegative ligand in carbanionic systems. Negative hyperconjugative delocalization becomes, however, less important as a driving force of agostic interactions in the case of electron-rich late transition metal complexes where M → L π back donation processes dominate and trigger the establishment of more covalent M···H–C interactions. We have therefore investigated the agostic phenomenon in early d^0 and d^n ($n \geq 2$) transition metal alkyls by a variety of experimental and theoretical techniques, and a clear and quite different picture has emerged.

MO analyses show that the β-agostic phenomenon in d^0 complexes can be described basically by *one* molecular orbital which accounts for the hyperconjugative delocalization of the M–C_α bonding pair over the β-agostic alkyl backbone and the establishment of secondary M···H interactions. Hence, agostic stabilization in early d^0 transition metal alkyls might have little or no dependence on M ← H–C electron donation, but arises rather from the delocalization of the M–C bonding electrons. In contrast, β-agostic interactions in d^n ($n \geq 2$) configurated alkyl complexes can be described in terms of an adopted Dewar–Chatt–Duncanson model. This model suggests three bonding components in case of our d^8 nickel alkyl reference systems: (1) Ni ← L σ donation, (2) Ni ← L π donation, and (3) Ni → L π back donation [26]. Hence, only in the case of late transition metal complexes covalent M···H–C interactions appear to play a dominant role, which are accompanied by characteristic features such as significantly activated C–H bonds and a pronounced lability toward β-H elimination. However, since electron delocalization of the M–C_α bonding pair still plays an important role for the energetic stabilization of agostic alkyl groups, the agostic interaction in d^n ($n \geq 2$) alkyl or amido complexes should be discriminated from the electronic situation in σ(X–H) complexes (X = C, Si).

In general, the structural parameters and charge density properties of agostic moieties in early d^0 and late d^n complexes differ remarkably due to the more covalent character of the M···H bonding in the latter case. However, the extent of the agostic interactions can be fine-tuned in all systems by the presence/absence of *local Lewis-acidic sites* at the metal atoms in the coordination area of the agostic C–H moieties. This is elegantly and clearly reflected by the correlation between the ^{1}H NMR shifts and the extent of the C–H bond activation with the local polarization pattern of the electron density at the metal [26, 90]. Large upfield shifts and highly activated C_β–H_β bonds are only observed in cases where the agostic hydrogen atom is pointing directly to a local Lewis-acidic center (charge depletion zone) at the metal atom. These local Lewis-acidic sites can be identified and characterized by charge density studies by visualizing and analyzing the fine structure of the Laplacian of the valence shell of charge concentration at the metal. In contrast to the original suggestion by BG, it is therefore not the *global* Lewis acidity of the metal center, but the presence of *locally induced sites* of increased Lewis acidity which determines the extent of agostic interactions [10, 11]. As a matter of fact, no agostic complex has been reported so far where agostic $C_\beta H_\beta$ moieties are pointing to a domain of local charge concentration at the metal. The nature and occurrence of all these LICCs can be derived by a direct and rigorous interpretation of the wavefunction: *they are an integral part of the bonds formed using metal orbitals possessing d-character. It is then the different nodal structures of p and d wavefunctions that give rise to different atomic polarization patterns at main group atoms and transition metals* [100].

To conclude: an understanding of the way in which the ancillary ligands induce polarization at the metal center, and of the interplay between these effects and the metal-alkyl bonding, affords the possibility of predicting – and hence controlling and manipulating – the development of an agostic interaction with an alkyl ligand in a particular situation. Such predictive power is unprecedented in this area of chemistry. Detailed analysis of molecular charge distributions thus offers the prospect of significant advances in the design and chemical control of complexes with central relevance to many reactions of academic and commercial importance [10].

Acknowledgments This work was supported by the DFG (SPP1178) and NanoCat (an international Graduate Program within the Elitenetzwerk Bayern)

References

1. Burawoy A (1945) Nature (Lond) 155:269
2. Pitzer KS, Gutowsky HS (1946) J Am Chem Soc 68:2204
3. La Placa SJ, Ibers JA (1965) Inorg Chem 4:778
4. Brookhart M, Green MLH (1983) J Organomet Chem 250:395
5. Trofimenko S (1967) J Am Chem Soc 89:6288
6. Trofimenko S (1968) J Am Chem Soc 90:4754
7. Trofimenko S (1970) Inorg Chem 9:2493
8. Brookhart M, Green MLH, Wong L-L (1988) Prog Inorg Chem 36:1
9. Scheins S, Messerschmidt M, Gembicky M, Pitak M, Volkov A, Coppens P, Harvey BG, Turpin GC, Arif AM, Ernst RD (2009) J Am Chem Soc 131:6154
10. Scherer W, McGrady GS (2004) Angew Chem Int Ed 43:1782
11. Scherer W, Sirsch P, Shorokhov D, Tafipolsky M, McGrady GS, Gullo E (2003) Chem Eur J 9:6057
12. Scherer W, Priermeier T, Haaland A, Volden HV, McGrady GS, Downs AJ, Boese R, Bläser D (1998) Organometallics 17:4406
13. Scherer W, Hieringer W, Spiegler M, Sirsch P, McGrady GS, Downs AJ, Haaland A, Pedersen B (1998) Chem Commun:2471
14. McGrady GS, Downs AJ, Haaland A, Scherer W, McKean DC (1997) J Chem Soc Chem Commun 1997:1547
15. Haaland A, Scherer W, Ruud K, McGrady GS, Downs AJ, Swang O (1998) J Am Chem Soc 120:3762
16. McKean DC, McGrady GS, Downs AJ, Scherer W, Haaland A (2001) Phys Chem Chem Phys 3:2781
17. Klooster WT, Brammer L, Schaverien CJ, Budzelaar PM (1999) J Am Chem Soc 121:1381
18. Klooster WT, Lu RS, Anwander R, Evans WJ, Koetzle TF, Bau T (1998) Angew Chem Int Ed Engl 37:1268
19. Cotton FA, LaCour T, Stanislowski AG (1974) J Am Chem Soc 96:754
20. Zerger R, Rhine W, Stucky G (1974) J Am Chem Soc 96:6048
21. Kaufmann E, Raghavachari K, Reed AE, Schleyer PvR (1988) Organometallics 7:1597
22. Dawoodi Z, Green MLH, Mtetwa VSB, Prout K (1982) J Chem Soc Chem Commun:1410
23. Dawoodi Z, Green MLH, Mtetwa VSB, Prout K, Schultz AJ, Williams JM, Koetzle TF (1986) J Chem Soc Dalton Trans:1629
24. Crabtree RH, Hamilton DG (1988) Adv Organomet Chem 28:299
25. Conroy-Lewis FM, Mole L, Redhouse AD, Litster SA, Spencer JL (1991) J Chem Soc Chem Commun:1601
26. Scherer W, Herz V, Brück A, Hauf Ch, Reiner F, Altmannshofer S, Leusser D, Stalke D (2011) Angew Chem Int Ed 50:2845
27. Calvert RB, Shapley JRJ (1978) Am Chem Soc 100:7726
28. Okuda J, Eberle T, Spaniol TP (1997) Chem Ber 130:209
29. Sinnema P-J, van der Veen L, Spek AL, Veldman N, Teuben JH (1997) Organometallics 16:4245
30. Carr N, Dunne BJ, Mole L, Orpen AG, Spencer JL (1991) J Chem Soc Dalton Trans:863
31. ADF2009.01, SCM, Theoretical Chemistry. Vrije Universiteit, Amsterdam. http://www.scm.com
32. Te Velde G, Bickelhaupt FM, van Gisbergen SJA, Fonseca Guerra C, Baerends EJ, Snijders JG, Ziegler T (2001) J Comput Chem 22:931, and references quoted therein
33. Becke AD (1988) Phys Rev A 38:3098
34. Perdew JP (1986) Phys Rev B 33:8822
35. Krykunov M, Ziegler T, Lenthe E (2009) J Phys Chem A 113:11495
36. Bader RFW, Matta C (2004) Organometallics 23:6253
37. Bader RFW (1998) J Phys Chem 102:7314

38. Popelier PLA, Logothetis G (1998) J Organomet Chem 555:101
39. Bader RFW, Tal Y, Anderson SG, Nguyen-Dang TT (1980) Isr J Chem 19:8
40. Tal Y, Bader RFW, Nguyen-Dang TT, Ojha M, Anderson SG (1981) J Chem Phys 74:5162
41. Pantazis DA, McGrady JE, Besora M, Maseras F, Etienne M (2008) Organometallics 27:1128
42. Etienne M, McGrady JE, Maseras F (2009) Coord Chem Rev 253:635
43. Thakur TS, Desiraju GR (2007) Theochem 810:143
44. Vidal I, Melchor S, Alkorta I, Elguero J, Sundberg MR, Dobado JA (2006) Organometallics 25:5638
45. Poater J, Solà M, Duran M, Fradera X (2002) Theor Chem Acta 107:362
46. Gatti C, Lasi D (2007) Faraday Discuss 135:55
47. Abramov YA (1997) Acta Crystallogr A53:264
48. Macchi P, Sironi A (2007) In: Matta CF, Boyd RJ (eds) The quantum theory of atoms in molecules. Wiley, Weinheim, pp 364–367
49. McGrady GS, Sirsch P, Chatterton NP, Ostermann A, Gatti C, Altmannshofer S, Herz V, Eickerling G, Scherer W (2009) Inorg Chem 48:1588
50. Frisch MJ, Trucks GW, Schlegel HB, Scuseria GE, Robb MA, Cheeseman JR, Montgomery JA Jr, Vreven T, Kudin KN, Burant JC, Millam JM, Iyengar SS, Tomasi J, Barone V, Mennucci B, Cossi M, Scalmani G, Rega N, Petersson GA, Nakatsuji H, Hada M, Ehara M, Toyota K, Fukuda R, Hasegawa J, Ishida M, Nakajima T, Honda Y, Kitao O, Nakai H, Klene M, Li X, Knox JE, Hratchian HP, Cross JB, Bakken V, Adamo C, Jaramillo J, Gomperts R, Stratmann RE, Yazyev O, Austin AJ, Cammi R, Pomelli C, Ochterski JW, Ayala PY, Morokuma K, Voth GA, Salvador P, Dannenberg JJ, Zakrzewski VG, Dapprich S, Daniels AD, Strain MC, Farkas O, Malick DK, Rabuck AD, Raghavachari K, Foresman JB, Ortiz JV, Cui Q, Baboul AG, Clifford S, Cioslowski J, Stefanov BB, Liu G, Liashenko A, Piskorz P, Komaromi I, Martin RL, Fox DJ, Keith T, Al-Laham MA, Peng CY, Nanayakkara A, Challacombe M, Gill PMW, Johnson B, Chen W, Wong MW, Gonzalez C, Pople JA (2004) Gaussian 03 Revision B.03
51. Mole L, Spencer JL, Carr N, Orpen AG (1991) Organometallics 10:49
52. Brookhart M, Green MLH, Parkin G (2007) Proc Natl Acad Sci USA 104:6908
53. Mitoraj MP, Michalak A, Ziegler T (2009) Organometallics 28:3272
54. Scherer W, Sirsch P, Grosche M, Spiegle M, Mason SA, Gardiner MG (2001) Chem Commun:2072
55. Scherer W, Sirsch P, Shorokhov D, McGrady GS, Mason SA, Gardiner M (2002) Chem Eur J 8:2324
56. Scherer W, Eickerling G, Tafipolsky M, McGrady GS, Sirsch P, Chatterton NP (2006) Chem Commun:2986
57. Chatt J, Duncanson LA (1953) J Chem Soc:2939
58. Dewar MJS (1951) Bull Soc Chim Fr 18:C71
59. Macchi P, Proserpio DM, Sironi A (1998) J Am Chem Soc 120:1447
60. Scherer W, Eickerling G, Shorokhov D, Gullo E, McGrady GS, Sirsch P (2006) New J Chem 30:309
61. Frenking G, Fröhlich N (2000) Chem Rev 100:717
62. Overgaard J, Clausen HF, Platts JA, Iversen BB (2008) J Am Chem Soc 130:3834–3843
63. Sparkes HA, Brayshaw SK, Weller AS, Howard JAK (2008) Acta Cryst B64:550–557
64. Reisinger A, Trapp N, Krossing I, Altmannshofer S, Herz V, Presnitz M, Scherer W (2007) Angew Chem Int Ed 46:8295
65. Krapp A, Frenking G (2008) Angew Chem Int Ed 47:7796
66. Himmel D, Trapp N, Krossing I, Altmannshofer S, Herz V, Eickerling G, Scherer W (2008) Angew Chem Int Ed 47:7798
67. Farrugia LJ, Evans C, Lentz D, Roemer M (2009) J Am Chem Soc 131:1251
68. Eliel EL (1972) Angew Chem Int Ed Engl 11:739
69. Eliel EL, Wilen SH (1994) Stereochemistry of organic compounds. Wiley, New York
70. Roberts JD, Webb RL, McElhill EA (1950) J Am Chem Soc 72:408

71. Schleyer PvR, Kos AJ (1983) Tetrahedron 39:1141
72. Reed AE, Schleyer PvR (1990) J Am Chem Soc 112:1434
73. Bader RFW, Slee TS, Cremer D, Kraka E (1983) J Am Chem Soc 105:5061
74. Cremer D, Kraka E, Slee TS, Bader RFW, Lau CDH, Nguyen-Dang TT, MacDougall PJ (1983) J Am Chem Soc 105:5069
75. Eisenstein O, Jean Y (1985) J Am Chem Soc 107:1177
76. Clot E, Eisenstein O (2004) Struct Bonding 113:1
77. Kocher N, Leusser D, Murso A, Stalke D (2004) Chem Eur J 10:3622
78. Kocher N, Selinka C, Leusser D, Kost D, Kalikhman I, Stalke D (2004) Z Anorg Allg Chem 630:1777
79. Ott H, Pieper U, Leusser D, Flierler U, Henn J, Stalke D (2009) Angew Chem Int Ed 48:2978
80. Ott H, Däschlein C, Leusser D, Schildbach D, Seibel T, Stalke D, Strohmann C (2008) J Am Chem Soc 130:11901
81. Tafipolsky M, Scherer W, Öfele K, Artus G, Pedersen B, Herrmann WA, McGrady GS (2002) J Am Chem Soc 124:5865
82. Scherer W, Tafipolsky M, Öfele K (2008) Inorg Chim Acta 361:513
83. Cheeseman JR, Carroll MT, Bader RFW (1988) Chem Phys Lett 143:450
84. Hartwig JF (1996) J Am Chem Soc 118:7010
85. Tsai Y-C, Johnson MJA, Mindiola DJ, Cummins CC, Klooster WT, Koetzle TF (1999) J Am Chem Soc 121:10426
86. Cai H, Chen T, Wang X, Schultz AJ, Koetzle TF, Xue Z (2002) Chem Commun:230
87. Matas I, Cámpora J, Palma P, Álvarez E (2009) Organometallics 28:6515
88. Allen FH (2002) Acta Cryst B58:380
89. Pupi RM, Coalter JN, Petersen JL (1995) J Organomet Chem 497:17
90. Scherer W, Wolstenholme DJ, Herz V, Eickerling G, Brück A, Benndorf P, Roesky PW (2010) Angew Chem Int Ed 49:2242
91. Pillet S, Wu G, Kulsomphob V, Harvey BG, Ernst RD, Coppens P (2003) J Am Chem Soc 125:1937
92. Wolstenholme DJ, Traboulsee KT, Decken A, McGrady GS (2010) Organometallics 29:5769
93. Forster TD, Tuononen HM, Parvez M, Roesler R (2009) J Am Chem Soc 131:6689
94. Papasergio I, Raston CL, White AH (1983) J Chem Soc Chem Commun:1419
95. Armstrong DR, Mulvey, RE, Walker GT, Barr D, Snaith R, Clegg W, Reed D (1988) J Chem Soc Dalton Trans:617
96. Perrin L, Maron L, Eisenstein O (2003) Faraday Discuss 124:25
97. Perrin L, Maron L, Eisenstein O, Lappert MF (2003) New J Chem 27:121
98. Klimpel MG, Anwander R, Tafipolsky M, Scherer W (2001) Organometallics 20:3983
99. Eppinger J, Spiegler M, Hieringer W, Herrmann WA, Anwander R (2000) J Am Chem Soc 122:3080
100. McGrady GS, Haaland A, Verne HP, Volden HV, Downs AJ, Shorokhov D, Eickerling G, Scherer W (2005) Chem Eur J 11:4921
101. Bader RFW, MacDougall PJ, Lau CDH (1984) J Am Chem Soc 106:1594
102. Weinhold F, Landis CR (2001) Chem Educ Res Pract Eur 2:91
103. Bader RFW, Gillespie RJ, Martín F (1998) Chem Phys Lett 290:488
104. Bytheway I, Gillespie RJ, Tang T-H, Bader RFW (1995) Inorg Chem 34:2407
105. Gillespie RJ, Robinson EA (1996) Angew Chem Int Ed Engl 35:495
106. Fischer A, Tiana D, Scherer W, Batke K, Eickerling G, Svendsen H, Bindzus N, Iversen BB (2011) J Phys Chem A 115:13061
107. Adamo C, Barone V (1999) J Chem Phys 110:6158
108. Lein M (2009) Coord Chem Rev 253:625
109. Poater A, Solans-Monfort X, Clot E, Copéret Ch, Eisenstein O (2006) Dalton Trans:3077
110. Solans-Monfort X, Eisenstein O (2006) Polyhedron 25:339
111. Emsley JW, Feeney J, Sutcliffe LH (1966) High-resolution NMR spectroscopy, vol 2. Pergamon, Oxford, pp 825–826

112. Harney MB, Keaton RJ, Fettinger C, Sita LR (2006) J Am Chem Soc 128:3420
113. Sauriol F, Sonnenberg JF, Chadder SJ, Dunlop-Brière AF, Baird MC, Budzelaar PHM (2010) J Am Chem Soc 132:13357
114. Ruiz-Morales Y, Schreckenbach G, Ziegler T (1996) Organometallics 15:3920
115. Gatti C, Ponti A, Gamba A, Pagani G (1992) J Am Chem Soc 114:8634

Index